TIBETAN V

Tibetan Venture

André Guibaut

HONG KONG OXFORD NEW YORK
OXFORD UNIVERSITY PRESS
1987

Oxford University Press

Oxford New York Toronto
Petaling Jaya Singapore Hong Kong Tokyo
Delhi Bombay Calcutta Madras Karachi
Nairobi Dar es Salaam Cape Town
Melbourne Auckland

and associated companies in
Beirut Berlin Ibadan Nicosia

First published by John Murray (Publishers) Ltd. London 1947
This edition reprinted, with permission and
with the addition of an Introduction, in
Oxford Paperbacks 1987

ISBN 0 19 584214 6

Printed in Hong Kong by Liang Yu Printing Fty. Ltd.
Published by Oxford University Press, Warwick House, Hong Kong

INTRODUCTION

SINCE the middle ages mystery has lain at the heart of Tibet's appeal to the West. Remote and inaccessible behind its mountain barriers, Tibet (Xizang) was not even a name to Europeans until the early fourteenth century. It was then that a Franciscan, Odorico of Porduone, first described it, but it is doubtful whether he went there, and it was not until the German Jesuit, John Grueber, reached Lhasa in 1661 that Europe was to receive any accurate knowledge of the country. The Jesuits were drawn to Tibet by the medieval legends of lost Christian communities. Their missionaries failed to find any but discovered instead many parallels between Christianity and Lamaism and the compelling spiritual power of the country.

Other Europeans followed them in pursuit of trade, gold, or political influence, but by the nineteenth century Tibet, then under Chinese domination, was expelling any European who tried to penetrate its fastness. Lhasa became 'the forbidden city', and to the mystery was added the challenge not only of surviving the perils of the journey and the harshness of the climate, but also of near-certain detection and expulsion long before Lhasa could be reached. While Western gunboats opened up the harbours of Japan and subdued even the Chinese Empire, Tibet stood aloof and fortress-like, an affront to every Westerner who in the name of science, religion, or progress sought to violate its frontiers. Of those who tried, a heavy toll was exacted. It was as if the mysterious power of Tibet were protecting its own. As André Guibaut was to write many years later in *Tibetan Venture* of the disasters which befell him, he became fearful of 'some malevolent force whose aim was to preserve from our impious prying eyes the last unknown territory of Central Asia'.

The expedition of Guibaut and Liotard, which is the subject of *Tibetan Venture*, has therefore a place in a long, if tenuous, history of European exploration on the roof of the world. It was their second expedition in Tibet together. The first, of which Guibaut was to write a much later account in *Missions Perdues au Tibet* (Paris, André Bonne, 1967), took place in 1936–7 and explored the middle reaches of the Valley of the Salouen. He explains in the preface to *Tibetan Venture* how in their hurry to return to Tibet he and Liotard delayed writing a full account of the first expedition. Then, after the long absence forced on him by the Second World War, Guibaut's first thought on returning to France in 1945 was to write the book 'consecrated to a journey which cost my friend his life'.

Tibetan Venture is therefore far more than a travel book about a remote country, or a dry account of a scientific and geographic expedition. The object of Guibaut and Liotard's journey was to advance anthropological knowledge of the Ngolo-Setas tribes of herdsmen in Eastern Tibet, to discover and map the source of the Tong (Tatu), and to demarcate the basins of the Blue and Yellow Rivers. But, written from notes five years after the events it describes, the book gained from that process in the writer's mind by which experience becomes also art, a process described by Wordsworth as 'emotion recollected in tranquillity'. The result is a book of acute observation and also of an inner tragic vision. The writing suggests by its intensity Guibaut's cathartic reliving of the events which appeared to lead inexorably and fatefully to his friend's death.

That tragedy was to be for him part of the wider catastrophe of the Second World War and the fall of his beloved France. His awareness of the perils facing civilization, of which the 'barbaric age of humanity' still preserved in Eastern Tibet remained ignorant, gave him an acute vision of how fragile was the world he was describing, and therefore of the responsibility which was his as possibly one of the last observers of a 'very ancient epoch of mankind'. To the modern reader the ominous horsemen silhouetted against the sky, who appear and reappear in the story, bring to mind the horsemen of the Apocalypse and so link Liotard's fate with the tragic destiny so soon to befall Tibet. They also give dramatic shape and tension to the narrative which reaches a climax in the ambush on the high pass, the gun fight, the last single shot, and then 'nothing but the great silence of high mountains' and 'a white horse grazing in the brushwood' which Liotard was never to reclaim.

Liotard's death ended a friendship of five years which had survived months of arduous travel in which they were alternately scorched, drenched, and frozen by the harsh Tibetan elements. In 1940 Guibaut was 37 and Liotard 36. Before their first journey to Tibet they had both served in the French merchant marine. Guibaut, who was born and educated at Bordeaux, qualified as a pilot before he became a sailor and then an officer in the merchant marine. He later studied at the Institute of Ethnology, acquiring the knowledge which he was to put to use in his Tibetan expeditions and learned publications. He refers briefly to his early life as a sailor in *Tibetan Venture* when he writes of 'other nights spent in the open, stretched out at full length on the deck in the tropics'.

His friend, Louis Liotard, was the son of the explorer and French colonial governor, Victor Liotard. He therefore had his father's example before him when he decided to join Guibaut in exploring one of the few last unmapped regions of the world. After Liotard's death Guibaut returned to China to begin a new life as a diplomat representing De Gaulle and the Free French in Chungking (Chongqing). After the war he became the French consul-general at Singapore, Casablanca, and Milan, and then ambassador, first to Jordan and then to Ceylon (Sri Lanka). He ended his diplomatic career as high commissioner in Laos and ultimately became the Director of the Office for the Protection of Refugees, in Paris. He died in 1966. He was a founding member of the French Society of Explorers and the winner of several medals and prizes for his geographical and scientific work. His translator, Lord Sudley, whose skill contributed to the success of the English version — it was chosen as Book of the Month by the Readers' Union — was to become well known in England as a journalist under his later title of Earl of Arran.

There was a tradition of French missionaries and explorers in Tibet whom Guibaut and Liotard were conscious of following. Guibaut mentions the Lazarist priest, Evarist Huc, who with another Lazarist, Joseph Gabet, reached Lhasa in 1846. Prince Henri-Philippe of Orléans had been a distinguished explorer of Tibet at the end of the nineteenth century. There was also the ill-fated expedition led by the former French naval officer, Jules Dutreuil de Rhins, who was killed after being turned back from Lhasa in 1893. It was his fate which seemed to haunt Guibaut as he and Liotard set out on their exploration, like him from Tat-sienlou (now Kangding).

What is striking to the English reader accustomed to a more phlegmatic national tradition is the degree to which Guibaut identified himself and his work with the civilizing mission of the French nation. Partly this was emotional compensation for the disasters which befell France in 1940, but it was also a distinguishing strand in the impulse which drove Frenchmen to explore and to colonize. 'Whatever success may be granted to our mission,' he wrote at the outset, 'we shall be working for that French community, by adding to its spiritual heritage if we succeed, and by offering our exertions, perhaps even our lives, if we fail.' There was some comfort even to be extracted from Liotard's death, as he explained in the Dedication, that 'another French, truly French, name will henceforward take its place on the map of the globe'. Liotard's name, and that of a French missionary had been added to the 'roll of French martyrs in Tibet. One of these men had died for his faith, the other for science, both for a disinterested cause and for the greatness of their country.'

By the time of Guibaut and Liotard's first expedition in 1936 Tibet's independence had been undermined by the expansionist ambitions first of China, then of Russia and of British India. The

Chinese first made their claim to suzerainty effective when their army invaded Tibet in 1720 to repel a Mongol force which had conspired with a faction of the lamas to take over the government in Lhasa. The Chinese reinstated the seventh Dalai Lama, established their own garrison in Lhasa and installed political residents, or *ambans*, to control the Tibetan government. It was under their unpopular regime that the Tibetans were forced to close their frontiers to the West. Before that they had shown European missionaries a friendly tolerance.

As the authority of the Manchu emperors crumbled in the nineteenth century before the pressure of the Western powers, so Chinese authority diminished in Tibet. After China's defeat by Japan in 1895 Peking's overlordship became merely nominal, but by then both Russia and British India were seeking to bring Tibet under their own influence as a buffer state. Russia's aim was the domination of Central Asia, an ambition which made British India fearful for the security of its northern frontier. It was this fear which led the Government of India to send Francis Younghusband with a military expedition to Tibet in 1903. After some bloodshed it forced its way to Lhasa and finally ended Tibet's isolation from the West. The strange power of the country, though, made a deep impression on Younghusband. After his entry into Lhasa he felt moved by a spiritual experience of an intensity he had never known before, which left him with a profound sense of well-being and contentment.

However, both the British and the Russians had underestimated the tenacious imperialism of the Chinese. In Sinkiang (Xinjiang), the Chinese province to the north, they had shown in 1877 that they never abandoned territory which they had once ruled. Similarly in 1909 they once more invaded Tibet to reassert their suzerainty. It was only the outbreak of the Chinese revolution in 1911 and the deposition of the Manchu emperors which gave the Tibetans the chance to rise against the occupying Chinese garrisons and allowed the Dalai Lama, who had fled to India, to return to rule his country.

In the 1930s Chinese authority was a reality only in the part of Eastern Tibet which was made into the new Chinese province of Xizang. The governor of the province was one of the old war-lords, or generals, who had carved up China between them after the overthrow of the Empire. However, under the leadership of Chiang Kai-chek the Nationalists had taken Peking in 1928. Despite continued warfare with provincial armies, and after 1931 against the Japanese, Chiang's writ still ran in Xizang. Guibaut noted that the provincial war-lord had turned into a civil servant, and that, even in such a drastic reform as the suppression of opium-smoking, the Nationalist government had made some headway in this remote province. None the less its people and culture remained Tibetan, and between the Chinese settlements in Xizang and the area of Central Tibet governed from Lhasa, which the Chinese still looked upon as a vassal state, there were large areas of country which knew no kind of authority at all. Such were the lands of the Ngolo-Setas.

Guibaut was favourably impressed by Chinese rule in Eastern Tibet. He described it as intelligent and tolerant, well adapted to a people who had shown themselves throughout their history as the most stubborn in resisting assimilation. Its mild, lazy appearance, lacking in obvious military display, its reliance on a system of rough, summary justice and also of 'gentle persuasion' was calculated to appeal to the self-interest of the Tibetans, and to reconcile them by trade and profit to Chinese rule. He had smiled at the label 'barbarian' which the Chinese gave to the nomadic population of the province, but the image which came to his mind when he observed the Chinese in their soft, silken robes jostling with the sheepskin-clad caravaneers of Central Asia was that of 'ancient Rome, in some emporium on the fringes of the Empire where the togas of the citizens mingled with the bear-skins of the barbarians'.

The tragedy of Liotard's death in an ambush by Ngolo brigands inevitably sharpened his sense of the benefits of Chinese civilization. His 40 days spent alone with his two Tibetan servants in constant danger of death, living with dirt and vermin, had conjured in his mind not only fantasies of French suburban comfort, but also an awareness that even humble Chinese workmen were by contrast with the Ngolo-Setas 'men who also had a long civilization behind them'. Even a squalid camp of Chinese goldwashers restored to him a sense of security, 'the sense of lawfulness as opposed to the law of the jungle, of despotism and fantasy run riot'.

The Ngolo-Setas tribesmen belonged to that other Tibet which few Europeans had penetrated, the Tibet of the primitive herdsmen of the high plateaux far removed from the civilization of Lhasa and of the settlements of the valleys. In that vast open landscape at an average height of 12,000 feet, beautiful in summer with thick waving grass festooned with flowers, man was the greatest enemy of man. Unseen eyes might observe the traveller and plan an ambush days in advance. Every ridge, every pass was to be feared as the possible site of an attack.

Guibaut felt no hatred, though, for the tribesmen who killed his friend. They were part of a world ruled by superstition, and fear of deities and fates. Their ethics, he observed, 'only take account of particular situations and . . . have no intrinsic conception of right and wrong'. The same tribesmen, in the person of one who accompanied Guibaut back to civilization, showed a spontaneous tenderness and a willingness to fight to protect him which astonished him with 'the contradictory nature of this extraordinary people, rough and yet gentle, one of the most attractive people in the world . . . who so strangely combined brutality with proofs of kindliness and good nature'.

In the settlers of the valleys, whose lives were focused on their monasteries, Guibaut found natural friendliness and in the lamas in particular an innate, simple kindness and compassion. His experience confirmed the picture given by Thubten Jigme Norbu, the brother of the Dalai Lama, of a people who found in their religion a way of happiness. The very beauty of the valleys, luxuriant in summer with lilacs, wild roses, and dense fields of barley, directed their minds beyond themselves, 'but the greatest beauty is that the people live a life dedicated to religion. You know it when you meet them, without being told. There is a warmth that touches you, a power that fills you with new strength, a peace that is gentle.'*

This was certainly what Guibaut experienced in the monastery of Dekho, despite the squalor of his surroundings, in the agonizing days after Liotard's death. Buddhism had come to Tibet in the sixth century preaching 'the Middle Way' as the path to a life of right living, of compassion and contentment. It was a tolerant religion which enabled every monastery to reserve a place for the primitive observances of Tantrism. Guibaut wrote about these: 'Everything that the extravagance of oriental wizardry would possibly conceive, everything that the disordered minds of hermits could invent is to be found in this luxuriant ritualistic display, in which cruelty and eroticism are interwoven.' He could only explain this extraordinary combination of religions by man's awareness in the thin air on the roof of the world of his closeness to the unseen and to nameless terrors against which the only resort was prayer.

He himself was conscious that the altitude, and his moral and mental isolation from the outside world, blotted out for him conscience and memory and everything but the instinct of self-preservation. This could lead either to 'a kind of sublime detachment, based on forgetfulness, which comes very near to absolute bliss', or to a fatalism, a conviction of a malevolent presence, which he was to describe as 'the baneful influence of the spell of Tibet'. After Liotard's death he experienced the further terror of being left alone unable to escape from the encircling mountains of a vast continent.

Guibaut's ability to analyse his own emotional and psychological responses as well as his power to bring before us a people and a way of life which has now disappeared, gives his book an appeal which transcends the differences between East and West.

---

* Thubten Jigme Norbu and Colin M. Turnbull, *Tibet: Its History, Religion and People* (London, Chatto and Windus, 1968), p. 345.

He had envisaged that the civilization of the lamas would dissolve gradually under the impact of tourism. He did not foresee how quickly and how brutally the world he described in 1945 would be destroyed by the Chinese nation he had grown to love. In 1950 the troops of the new Communist government moved into Tibet. By the time that the Red Guards had completed the programme of Cultural Revolution there were few monks left in a population which had formerly given almost one in three of its menfolk to the monasteries. Most of the monasteries, the ancient shrines, and much of Tibetan art had been destroyed. The children of a once-happy people were compelled to substitute atheism and the class war for the quiet contentment of the Middle Way.

It is this tragedy of the destruction of an ancient civilization and of the soul of a people which causes us to value afresh *Tibetan Venture*. For it was Guibaut's own personal tragedy which enabled him to experience and write movingly of the common humanity and goodness of men who belonged to a world so distant from his own and which was so soon to perish. Differences of language, nationality, and culture ceased to be significant in the 'kind of distant echo of a mother's affection' which he received from the lamas of Dekho, which is one of the many abiding memories of this book.

PAMELA NIGHTINGALE

Tibetan Venture

In the Country of the Ngolo-Setas

SECOND GUIBAUT-LIOTARD EXPEDITION

by

ANDRE GUIBAUT

Translated by LORD SUDLEY

LONDON

JOHN MURRAY, ALBEMARLE STREET, W.

To LOUIS VICTOR LIOTARD

Explorer

My companion
killed by Tibetan bandits
September 10, 1940
on a pass which to-day bears his name

latitude north 32° 21′
and
longitude east 100° 24′
in the country of the Ngolo-Setas

BORN in Paris, *April* 9, 1904, son of Victor Liotard, Colonial Governor, Explorer and Colonizer of the Haut Oubangui, and of Madame Liotard, *née* Lachave. Lieutenant in the Merchant Navy.

First Guibaut-Liotard mission (1936–37): to explore the valley of the middle reaches of the Salouen.

Second Guibaut-Liotard mission (1940): to explore the territory of the Ngolo-Setas and the upper basin of the Tong.

Holder of the Gold Medal of the Geographical Society of Paris, Louise Bourbonnaud prize—of the Medal of the Société de Géographie Commerciale, Henri d'Orleans prize.

Lauréat de l'Institut.

Died on mission to Tibet, *September* 10, 1940.

Chevalier de la Légion d'Honneur (Posthumous).

Your death, following the death of so many others, in 1940, when France seemed to be faced on all sides by disaster, passed almost unnoticed. Nevertheless you too gave your life for your country and for mankind. You were killed in the middle of one of the last white spots on the world's atlas, in the vanguard of civilization, facing the last of the barbarians. Your tomb, the highest in the world, the nearest to Heaven, is one of the signposts that mark the successive and often painful stages of the earth's discovery. It is a point—mathematically exact in the geodesic sense—the beginning of a road. It is no blind alley. Others will tread this road. They will go further; they will reach the goal we hoped for. They too, perhaps, will die, marking with their bodies a further stage of human progress. Millions more are giving their lives for that progress to which, in spite of everything, we must pin our faith. Thanks to you another French, truly French, name will henceforward take its place on the map of the globe.

(Written in exile in Chengtu, China. *September* 10, 1941.)

I earnestly beg my Chinese friends, and in particular the scholars of the "Academia Sinica" and of the "Service Géologique de Chine", to whom I am bound by long years of friendship, to preserve on their maps the name of Liotard with which, in token of respect, I ventured to christen the mountain pass where my companion met his death. I thank them warmly, and send them the expression of my deepest affection.

CONTENTS

LIST OF ILLUSTRATIONS

Between pages 22 and 23

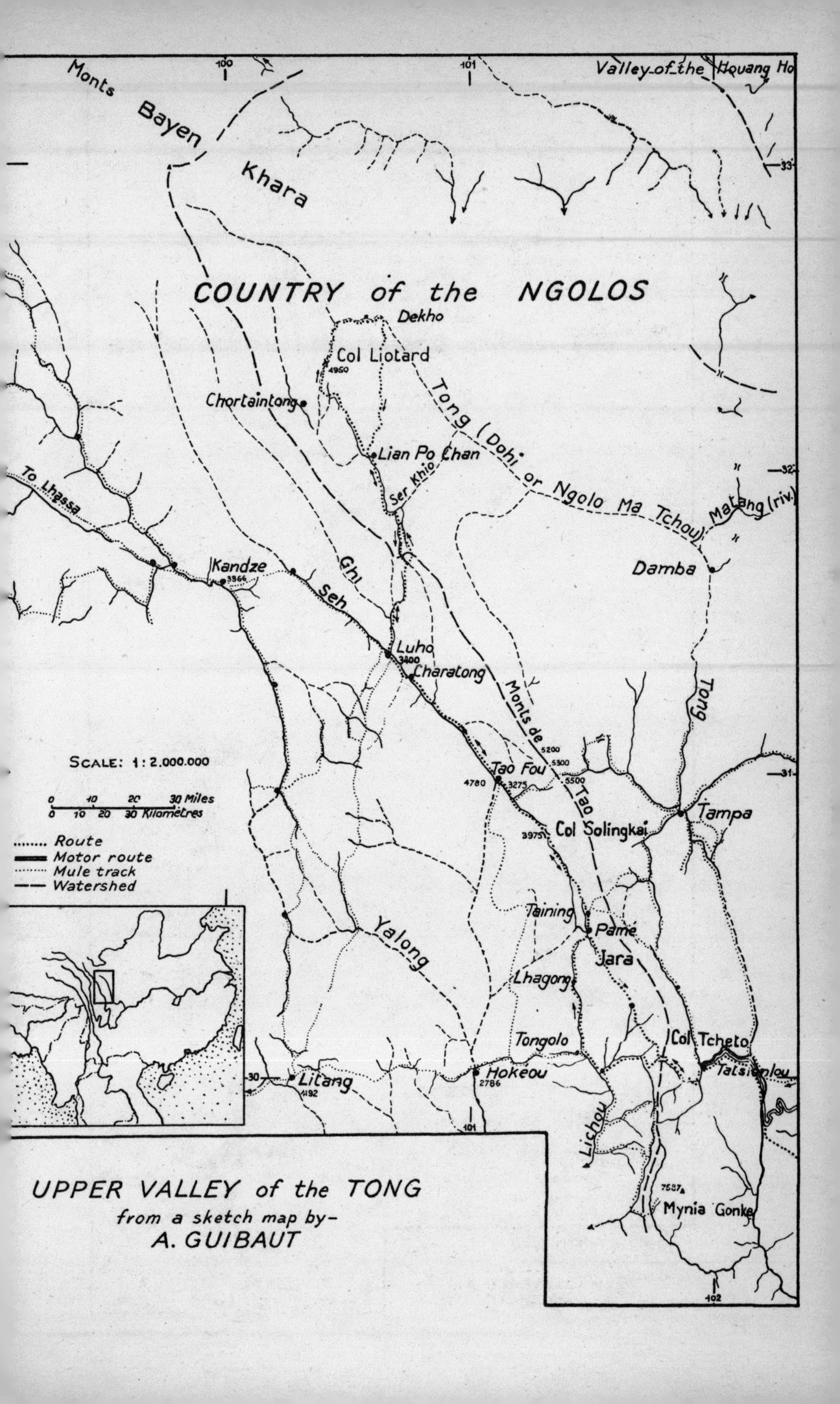

Monts Bayen Khara
Valley of the Houang Ho
100
101
33
32
31
30
102
COUNTRY of the NGOLOS
Dekho
Col Liotard
4950
Chortaintong
Lian Po Chan
Ser Khio
Tong (Dohi or Ngolo Ma Tchou)
Matang (riv.)
Damba
To Lhassa
Kandze
3866
Ghi
Seh
Luho
3400
Charatong
Monts de Tao
Tong
5200
5300
5500
Tao Fou
4780
3275
Tampa
3975
Col Solingkai
SCALE: 1:2.000.000
0 10 20 30 Miles
0 10 20 30 Kilometres
Route
Motor route
Mule track
Watershed
Taining
Pamé
Jara
Yalong
Lhagong
Tongolo
Col Tcheto
Tatsienlou
Litang
4192
Hokéou
2786
Lichou
7587
Mynia Gonka
UPPER VALLEY of the TONG
from a sketch map by–
A. GUIBAUT

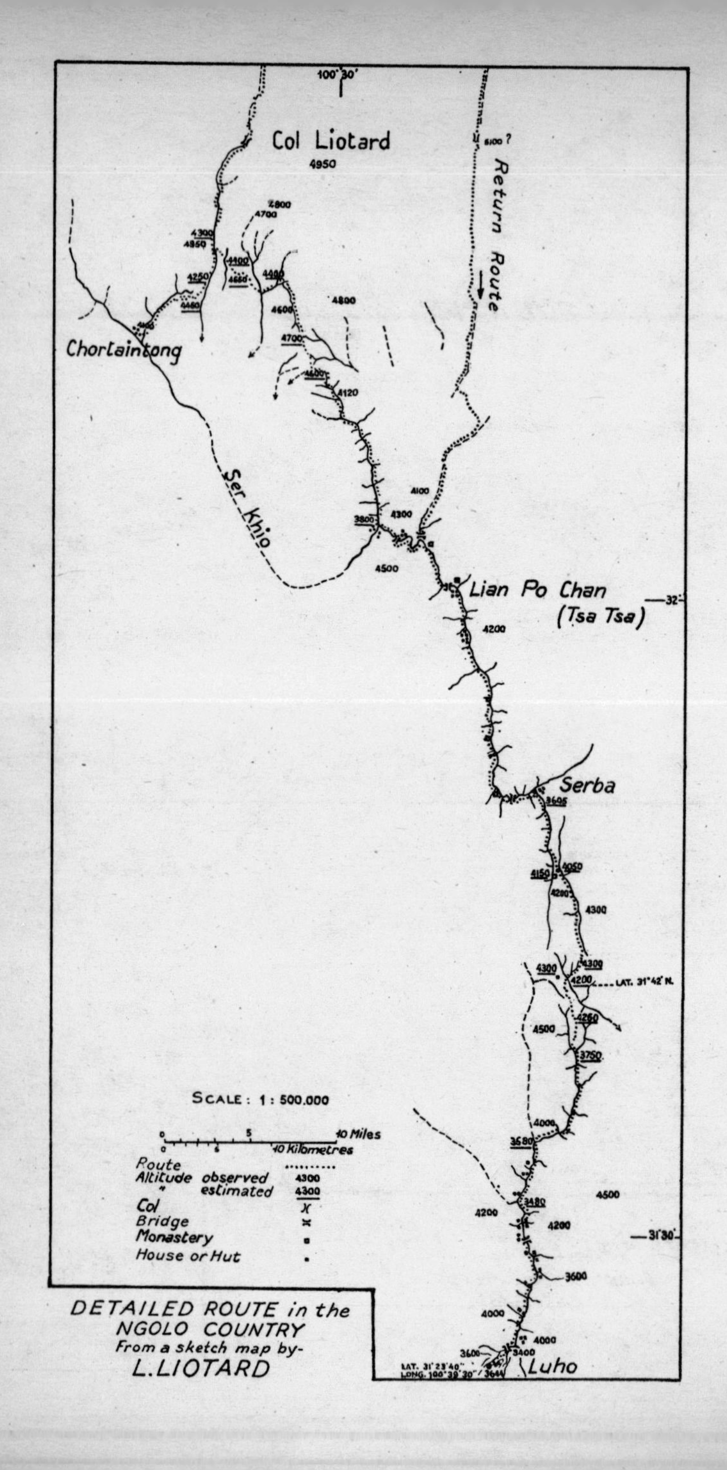

DETAILED ROUTE in the NGOLO COUNTRY
From a sketch map by-
L. LIOTARD

List of Illustrations

MAPS

The Koukou-Chili or Bayen-Khara Mountains divide the Blue River from the Houang-Ho. We noticed as we approached in a north-south orientation that these ranges changed their geological constitution, and became at this point analagous to the parallel undulations which align themselves in the north-east along and behind the borders of the Sseu-tchouan. It is not known how these two bearings meet, for the territory where they converge, southwards from the meridional belt of the Houang-Ho, is the least known territory in Asia and in the whole world.

Fernand Grenard

(GÉOGRAPHIE UNIVERSELLE, VOLUME VIII)

PREFACE

It is scarcely credible that there are people who are still ignorant, and will perhaps remain ignorant all their lives, of the catastrophe which has just befallen mankind. The black people of Africa, the nomads of the Semitic world, the Eskimos of Greenland, the Kayaks of Borneo, the Kachins of Burma, and even the Lissous of the Salouen whom Liotard and I were the first to discover, all these clans or tribes have been in one way or another affected by the war. Probably the herdsmen in the higher table-lands of Tibet, in the vast solitude of their surroundings, are the only people unaware that the world became insane, and who can still use the single word "Piling" (stranger) to denote all men who have not had the good fortune to be born Tibetans. Tibet is probably the only territory which has not been caught up in the world conflict.

Whilst other countries are being drawn closer together, while time and space are losing the permanent value which they have possessed for thousands of years, from the days when men could only estimate distances by day's marches, Tibet, far removed in time and knowledge from our now crazy civilizations, has not changed.

Any explorer, driven by the thirst for discovery, who has traversed the labyrinth of mountains, dark gorges and torrents of geological antiquity which lead up to these vast, almost desolate plateaux topped by glaciers reaching almost to the stratum of absolute cold, feels that he is advancing into a world of four dimensions. He progresses as far vertically as horizontally; he lives among people of a different era to his own, people who, from the fact that they are ignorant of certain things which have no reality for them, things which have been invented and created within the past four centuries, seem to have lost some faculty of understanding. Stupendous journey into Time!

But though Tibet, protected as much by its high mountains as by its boreal climate, has managed to escape for a few years from the tragic community of nations, it is not likely to preserve its anachronistic state much longer. Already men of strange purpose, armed with small wooden pegs and sighting apparatus have penetrated into its territory, and subsequently roads have bitten into the soil of the Lamas and airfields have flattened it.

The Tibetans do not understand these wooden pegs with which the Chinese engineers mark out their tracks, and which bear some resemblance to the pegs which the wild tribes of the eastern mountains, on the far fringes of Tibet, plant round their settlements to drive away evil spirits. They do not realize that these pegs are warning signs that a very ancient civilization, now condemned, is about to disappear. Will that which is to come be an improvement? The time is obviously near when it will be possible to penetrate into Tibet by car or plane. Then Lama civilization will dissolve into tourism. Perhaps I shall be one of the last to have known that very ancient epoch of mankind.

Thus, although it stirs in me a wealth of painful yet precious recollections, there is still time to recount the last Guibaut-Liotard mission, an expedition in which we travelled through Central Asia as Marco Polo, Père Huc and Gabriel Bonvallot travelled before us, by caravan, in the approved fashion. It is an expedition from which I returned alone, as Grenard had done more than fifty years ago, after the expedition which he undertook with Dutreuil de Rhins.

. . . . .

Those who have been kind enough to take an interest in the Guibaut-Liotard missions will doubtless be surprised that an account of the second expedition should appear before an account of the first one has been published. I therefore owe my readers a few words of explanation.

Preface

When we returned from our first expedition[1] Liotard and I decided to return at the first possible opportunity. We therefore purposely limited ourselves to the publication of a few articles in newspapers and reviews, intending to reserve for a future date the production of a single work which should give a combined account of the two journeys.

The declaration of war surprised us as we were about to make our departure. Since the political authorities were at that moment concerned with the propagation of Franco-Chinese relations, it was decided, at the suggestion of M. Mandel and General Buhrer, that, considering its purely scientific nature and the fact that the Chinese were friendly disposed towards it, our mission should be carried out in spite of the political situation. So we embarked at Marseilles on the *Athos*, on November 30, 1939, thinking that we should be returning in a year's time.

After the expedition and the death of my companion, I returned to China in November, 1940. I rejected the offer made me by the Vichy Ambassador to return to France by way of Indo-China, and wired immediately to General de Gaulle to offer him my services. The General did me the honour of accepting, and asked me to remain in China, where no doubt there would be need of people who knew the country. Indeed, shortly afterwards, I became the representative of the Fighting French with the Chinese Government, after the departure of the first representative, Professor Escarra.

[1] The first Guibaut-Liotard mission, patronized by the Geographical Institute of the University of Paris, by the Geographical Society and by the Musée de l'Homme, took place in 1936–37: to explore the middle reaches of the valley of the Salouen. (Gold Medals from the Geographical Society, Louise Bourbonnaud prize, and from the Society of Commercial Geography; Henri d'Orleans prize; Academy of Sciences prize).

For facts relating to this expedition see:

(a) *Harvard Journal of Asiatic Studies*, Volume 3, Nos. 3 and 4, *December*, 1938

(b) *Annales de Geographie*, No. 283, 1941.

(c) *Annales de Geographie*, No. 293, 1945.

Refer to the second expedition article by A. Guibaut, published in the *Geographical Review*, Volume XXXIV, No. 3. 1944. (American Geographical Society.)

Preface

Years passed in this country of China which offered me kindness and hospitality. I occupied the leisure hours afforded me by my duties there, and by the missions with which I was later entrusted in Algiers and the United States, to set down on paper the account of the second expedition. I could not deal with the first, as all my notes and papers had been left in Paris.

I returned to France in April, 1945, and am now publishing the account of the 1940 expedition. A sense of reverence which readers will understand forbade me to postpone further the publication of a book consecrated to a journey which cost my friend his life. It is a book which I owe to his memory as well as to those who mourn him, above all to his mother, whose courage should serve as an example to everyone.

I hope soon to be able to publish the account of our 1936–37 expedition. It will appear after a delay of nearly ten years. But I am certain that my readers will forgive me, realizing that I devoted a great deal of that period to the service of my country.

CHAPTER I

TATSIENLOU, EMPORIUM OF TIBET, MAY 1940

For the past hour flames have been raging in the quarter of the Catholic Mission. Fire, worst of all scourges in a town of wood, broke out at about seven in the evening in a house near the bridge which straddles one of Tatsienlou's two large mountain torrents. Like an angry dragon intent on devouring the town, the furnace hisses, growls, and stretches out its tentacular and lively flames towards those wretched hovels vividly illumined, inflammable as tinder. The bed of the fire throws a dazzling white light against a large cob wall which is still standing, but which has lost the house once coupled against it. The mountain torrent reflects great waves of fire which, as the eye progresses, shine with a translucent green. On both sides of the town the light streams up the hillsides, illuminating the pine-trees to the summit.

On the bank of the river, in the narrow, paved street, a terrified crowd bustles about in the utmost confusion, trapped between fire, water, and mountains. Soldiers, volunteers, try to limit the area of the destruction by abandoning an important part of the town to this Moloch. In desperation, the unhappy people who are thus left to their fate, struggle to drag out their meagre belongings from their houses already drenched by showers of sparks. Those still at a distance from the fire, seeing a horde of people rushing in and out of their homes, hastily breaking up the roofs of their houses and throwing them into the street, heaving their furniture out of their windows and tearing up their rafters and floor-boards, start lamenting and bemoaning the cruel fate which has decreed that these unfortunates should be sacrificed for the sake of the community. Black writhing figures with faces upturned towards the reddened sky stand outlined against the blaze. Chinese and Tibetans come and go, disappearing under piles of clothes and furniture. A herd of yaks, whose stable-yard is threatened, passes grunting along the street, beaten from behind by its herdsmen. Dogs, their hair bristling with terror, set up a howl of death.

Profiting by the confusion, thieves prowl around among the homeless ones, while soldiers, with their fingers on the triggers of their revolvers, attempt to catch them red-handed. Every now and then a shot rings out, and the uproar increases.

A curious little sideshow is a terrified cat; the scurrying feet around him cause him to jump into the water and start swimming across a narrow channel. Pathetic symbol of the general panic!

For the last half-hour we have been wondering whether the Mission would be caught up in the general disaster. Little flakes of fire are falling on the buildings, and the servants perched on the roof-tops are busy throwing them back into the street. We and the missionaries prepare to evacuate the stores, and to rescue everything that can be rescued.

At about ten o'clock the wind fortunately takes a different turn, and a providential rain begins to fall. The lurid glow of the fire will last all night, but the danger is averted.

For the next three days the victims of the disaster are obliged to remain in their ruined homes. No one is anxious to shelter people who are marked by the evil spirit of Fire.

Nevertheless, out of sheer kindness of heart and from long experience of communal life, the inhabitants of the town bring food to these wretches, who are protected by no form of insurance, and who have lost everything they possess.

We have stepped right back into the Middle Ages.

. . . . .

A few days previous to this, on *May* 9 to be precise, Liotard and I passed through the east gate of Tatsienlou. A typical old arched Chinese gate, with enormous iron-clad panels studded with square-headed nails. We are at our starting-point. The real journey does not begin until we leave this town.

And yet how many miles have we travelled since leaving Marseilles! The number itself would be impressive. Interpreted into stages, it is interesting from the point of view of the variation in transport: thirty odd days by boat from Marseilles to Haiphong—three days by train to Yunnan—a week by bus from Kunming[1] to Lushien on the Blue River—two days by steamer to Chungking—an hour by plane from Chungking to Chengtu—four days by rickshaw from Chengtu to Yatchow (Yahan)—and finally seven days' ride by litter to the Catholic Mission in Tatsienlou! We are at the end of the world—of our world—and on the fringe of a new one.

It is this very town of Tatsienlou, where Chinese and Tibetan civilizations intermingle, that so many travellers and missionaries have used as their starting-point. Jacques Bacot, whom we are pleased to call our Master, left it one morning on a voyage of discovery which gave him the material for the world's most beautiful travel story.[2] It is here that le Père Huc, nearly a hundred years ago, accompanied by M. Gabet, re-entered China from Tibet with those spicy jottings which put him in the front rank of great reporters. And it is also in Tatsienlou that Gabriel Bonvallot and Prince Henri d'Orleans ended their marvellous journey across Central Asia. Many Frenchmen have passed through this little town, and some of them have never returned, but have perished in the solitudes of the West. For the traveller, Tatsienlou is to Tibet what Tongourt or El-Golea were to the Sahara: a starting-point for those whom faith or the spirit of adventure have driven out into the unknown like caravels across the ocean.

For more than fifty years there have been Frenchmen in Tatsienlou, missionaries from the Société des Missions Etrangères, whose peaceful seminary was founded in Paris in the seventeenth century, in the old Rue du Bac; men who have undertaken the

[1] New name for Yunnanfou.

[2] *Tibet in Revolt.*

bold and ungrateful task of preaching the Gospel in Tibet. They have paid a heavy toll in human lives and suffering for the honour of being in the vanguard of Christianity.

We were overjoyed to see them again, and greeted them like old comrades-in-arms. How gladly we steeped ourselves once more in the atmosphere of evangelical outposts that we know so well, a gay, almost casual atmosphere which reeks more of an officers' mess than of a vestry!

There are four or five Fathers in the Tatsienlou district, bearded men, stalwart as the trees in their provinces whose accents they have preserved, still very French in spite of their exile from their mother country. Many of these priests have never been back, and many of them will never see France again. Every evening we would meet in the library which they use as a sort of common-room. I love these evenings spent sitting round the brazier and the smell of the charcoal smouldering in its copper grate, which evokes, as only the magic power of scent can evoke, recollections of a winter season spent in Bahang in 1936–37, in the company of Père Burdin and Père Bonnemin, at a height of 8,200 feet when we were entirely cut off from the world by snowstorms which blocked the passes. We chat now as we chatted then, lighting and relighting our pipes. Repeating ourselves now and then, we call to mind various travellers and missionaries whom we have known. We tell the same tales, the same anecdotes, which have been handed down to us traditionally from old and new, embellished slightly by the passage of time and always listened to with compliant attention. There are not many opportunities for renewing the repertoire, and when we have finished recounting the small events of the day, there is not much left to talk about. The world is so far off. Thus there is a sort of tacit agreement, old as the Missions themselves, whereby everyone is indulgent towards the other's stories, hoping that his indulgence will be repaid him in kind. But there is no need to indulge the passers-through, for missionaries, like their friends the sailors, with whom they have certain qualities in common, are often gifted story-tellers.

This time, unfortunately, a sense of urgency dispels the wonted calm of our talk. The war in Europe is in full swing. We receive faint echoes of the conflict through the Chinese paper published in Tatsienlou. Every morning Mgr. Valentin takes great pains to translate the names of the places mentioned in the day's communiqué. We listen incredulously as he reads out the names of towns the very mention of which seems absurd to us. We little group of Frenchmen are near breaking-point, suffering in sympathy with our unhappy country.

Mgr. Giraudeau, for there are two bishops in Tatsienlou, asks the same question twenty times a day. "Where are the Prussians now?" The old man fought under Bourbaki in the 1870 campaign. To him Germans are always Prussians, and he tends to confuse the three wars. Head of the "Missions Etrangères", he came to China in 1878, having previously been a Pontifical Zouave, at the time when missionaries wore a pig-tail and dressed in the ancient fashion, according to their rank in the priestly hierarchy. He never returned to France, and now, having handed over the control of the Mission to Mgr. Valentin, he is spending his last days in peace in the country of his adoption. Mgr. Giraudeau, nearly a hundred years old, carries his age very well, and always refers to his fifty-year-old servant, who has been with him since his youth, as "my young man".

The day after we arrived there was an addition to our little company in the shape of Père Doublet. He entered the court-yard of the Mission holding his horse by the bridle, dressed in a long black robe, a rifle slung over his shoulder. He had just done the journey, at present rather a dangerous one, from Taofou to Tatsienlou. He brought with him, to liven our evening talks, his stimulating zest for life and his tales of previous adventures. As he had just spent several years *en poste*, that is to say, quite alone, he had a great need for conversation.

. . . . .

This morning one of the Provincial Governor's henchmen came to announce that his master intended to call upon us. We immediately prepared the drawing-room of the Mission, dusting the large Chinese seats, arranging the curious assortment of articles which has accumulated with the years, *objets d'art* from old France which a strange fate has brought to this far-distant spot.

This is the first visit which Marshal Lieou Wen-Hui has paid to the Mission. With a due respect for etiquette he is returning the visit which we paid him in his modest yamen, seat of the Government of Sikang. This pro-consul is no ordinary civil servant. He is one of the last representatives of the caste of Chinese feudal lords whom the Westerners have pompously christened "War Lords". This man, still of tender age, has already led twenty-five years of strenuous life, years which the inhabitants of his native province of Sseu-tchouan are not likely to forget. Head of a powerful clan, counting on the support of tens of thousands of adherents, he was inspired by competition of interests and "face" to wage a seven years' civil war against his nephew, who incidentally is older than himself, the stakes being the possession of the whole province. The outside world has known little or nothing of this piece of family history, although the territory under dispute was more extensive than many European countries, and has a population of roughly fifty-five millions. It was a typical war of the Middle Ages, interrupted by negotiations which ruined the people, who were obliged to pay their taxes a hundred years in advance, payments for which they received acknowledgements at regular intervals. Whether or no it was because his enemy had a pet magician and never engaged in battle without first consulting him (this magician is still alive, and has gone into retirement in Chengtu) the war between the two Lieous ended in the defeat of our friend. It so happened that one Marshal Lieou entered the capital of the province through the east gate while the other Marshal Lieou was fleeing through the west gate, which, through some piece of negligence on the part of the conquering army, had been left unguarded while the city was being occupied. There was nothing for the victor to do but to reassure the wife of the vanquished who had remained behind that no harm would come either to her person or, more important still, to her property.

Since then, Lieou Wen-Hui, under pressure from Chiang-kai-chek, who might well figure in history as the Richelieu of China, has kept within the law. The feudal lord has become a civil servant. He still retains some of his independence but, with the passing of the years, he has acquired wisdom and is beginning to realize more and more clearly that the age of Great Captains has lapsed, and that provincial particularism has had its span. Now, governor of an unproductive and sparsely-inhabited province, he has resigned himself to his lot, on condition of being allowed to relax frequently in Chengtu where, faithful to his promise, his conqueror has respected his seigniorial rights. With the approach of old age, and perhaps influenced by the proximity of the many Tatsienlou monasteries, he has become very devout. It is said that the Lamas have a strong influence over him. In any case he has won the support of the high priests of Lamaism by making them members of the Provincial Council. In this way he has shown himself to be a shrewd politician.

His visit is attended with a certain pomp. He arrives at the Mission in a large sedan-chair borne by four servants wearing the old-fashioned livery, preceded by outriders to divide the crowd, and followed by his guardian angels. He brings his dogs with him, a sheep-dog of European extraction and a black greyhound from Turkestan, who, the moment they set foot inside the Mission House, start fighting with the missionaries' dog, which causes no little confusion.

Face to face, we bow ceremoniously. He has great distinction, this soldier of fortune, in his blue flowing silken robe with sleeves reaching so low that they completely cover his hands. He wears no insignia of rank. Entirely shaven, having neither hair nor beard, his face is without ornamentation of any kind, and owes its virile beauty to his prominent cheek-bones, wide nose, set lips and, above all, to the small, slanting eyes with which he boldly scrutinizes us.

At the drawing-room door we indulge in a bout of politeness

to determine who shall not be the first to cross the threshold. We all know well, and the Marshal better than anyone, that as a visitor he must take pride of place; but it is right to respect custom and to put up a show of resistance, whose duration varies with the individual. A too short resistance would be uncivil, a too prolonged one in bad taste.

I like the present-day Chinese manners which have replaced the traditional formalities. The old courtesy, so precise and stereotyped in form as to smother all originality and consequently all natural delicacy of feeling, is now only practised in outlying country districts. As for the procedure which consisted in bowing one's head right down to the ground, that went out with the Empire. There was a time when Mandarins who were received in audience by the Viceroy had to lie full length on their stomachs at his feet, while officials of equal rank returned the courtesy in the name of the high and mighty potentate who himself remained calmly seated on his throne.

While we all sit around and exchange, through the medium of Mgr. Valentin, who acts as interpreter, the customary banal remarks about the weather, the country, our journey and so on, I look curiously at the Marshal, seeking for traces of opium on his face. They say he smokes in moderation. An aristocrat's privilege, which has no serious consequences on the health provided it is not abused, the practice of opium smoking is nevertheless forbidden by the National Government, because it has become a real danger for China, chiefly in social and economic spheres. In many provinces already, thousands of acres of soil, which had been previously consecrated to the poppy, are now turned over to the cultivation of more useful products. Anyone acquainted with China will be surprised that the present government has managed to carry through such a drastic reform. It is not the least of its achievements. But there is still a great deal to be done. Here, in the Si-kang province, public smoking-houses and opium-traders abound. But we are so far away from the capital here! And yet the effects of government control are already noticeable. Moreover, the suppression of a habit so deeply rooted in the minds of the population is bound to have such drastic social and economic results that one cannot blame the great Marshal for contriving a transition period.

As I sit watching the Governor my thoughts revert to the former potentate of the district, the Tibetan King of Tatsienlou, who came yesterday to visit his old friend the bishop. He has been recently dispossessed of his lands, having suffered from one of the frontier fluctuations which are so common in this borderland district. The rough, simple Tibetan lord later sold a large part of his lands, and has since then apparently become resigned to being a person of no consequence.

There are many of them in these frontier districts, local princelings who have lost their kingdoms in this manner. These fallen feudal lords are hardly distinguishable from the rich farmers of the neighbourhood. Their houses are sometimes more spacious, their dress more elaborate, but were it not for the respect which their former subjects pay them they would pass unnoticed. Chinese politicians seem to think it more important to acquire the support of the great lamaseries than to lean on the temporal power of the tributary lords. Moreover, China regards these territories as an integral part of its Empire, so that in principle they are not allowed to benefit by any regime of exception. In practice it is quite another matter—the Chinese are much too intelligent and tolerant to impose on those millions of its population who are not of Han extraction the rigid laws of a code which they apply fairly loosely even among themselves.

The province of Si-kang was created a few years ago. Theoretically Chinese, it arbitrarily embraces almost the whole of Eastern Tibet, though no one knows the exact frontier-line which divides it from Central Tibet, which China regards as a vassal state, and which enjoys a certain degree of autonomy. In this maze of mountains and valleys, whose topography is not yet exactly determined on the maps, only the communicating roads, marked out by villages and monasteries, are effectively known. A tacit agreement between Tibetans and Chinese demarcates along these roads the points controlled by Lhassa and by Chungking. But between these big roads the areas of allegiance are much more vague.

In point of fact the Chinese govern this province in a lax and easy-going fashion. The soldiers and civil servants, whose duty it is to control Si-kang, are scanty in number. You meet a hundred or so of them on your journey along the large caravan roads, distributed among a small number of military outposts, entirely cut off one from the other. Thus, although it possesses a Chinese name, Eastern Tibet has remained true to itself and, in spite of its situation on the map which places it within a day's march of its powerful neighbour, it does not differ in any profound sense from the other districts of Tibet, which astound the traveller by their ethnological, linguistic, religious and cultural unity. Let me say in passing that, of all peoples who have surrounded and still surround the cradle of the Hans, the Tibetans are the most stubborn in resisting the process of assimilation to which all China's neighbours, swamped by her irresistible vitality, sooner or later succumb.

When it became the metropolis of a Chinese province Tatsienlou surrendered its Tibetan name. Marshall Lieou Wen-Hui's capital is now called Kangtring. Nevertheless, in the course of this book, I shall continue to call it Tatsienlou, because of the sentimental associations which this name arouses in the minds of French travellers. It is still a town of the Middle Ages, with its narrow, paved streets lined with booths of merchandise, its gates closed regularly every evening at the moment when the night-watchmen start their rounds, marking the passage of the hours with big blows on their gongs, as the "serenos" used to do in the provincial cities of Spain. Here Tibetans and Chinese barter their wares; long processions of tea-carriers pass at regular intervals along the mountain tracks which start in the Sseutchouen and end in the granaries of Tatsienlou, and the great Tibetan yak caravans, loaded with bales of raw wool, end their long journeys in its caravanserais. All and sundry, Tibetans and Chinese, spend their money in its little restaurants and buy manufactured knick-knacks in its shops. Its streets are enlivened by picturesque crowds, Chinamen in soft silken robes jostling against the caravaneers of Central Asia in their rough sheep-skin coats. Lamas in red cloaks, elegantly draped, assemble in the squares, chattering away to each other until some initiate summons them to perform certain rites.

His visit ended, the Governor, perched high in his sedan-chair, returns to his yamen through the swarming streets, preceded, like a Roman pro-consul, by lictors who thrust roughly aside the bowing Chinese and the rough Tibetan traders, the latter astonished to see a warrior carried like a woman instead of riding off on a proud steed. We are back in the days of ancient Rome, in some emporium on the fringes of the Empire, where the togas of the citizens mingled with the bear-skins of the barbarians.

. . . . .

Nevertheless Tatsienlou has already been affected by progress. A crude paddle-wheel, propelled by the magnificent roaring mountain torrent, the noise of which deafens the inhabitants, has provided the town with electricity, and has even made it possible for an intelligent Chinaman to show silent films three times a week. These cinema performances are so crowded that it is quite impossible to read the titles. In the near future there will even be a wireless transmitting station. A former Grenoble student, M. Tcheou, has come to supervise its installation. I imagine the inauguration ceremony will be something quite out of the ordinary. We met in Tatsienlou the famous French traveller, Madame Alexandra David-Neel, and her adopted son, the charming Lama Yongden. Unfortunately, a few days after her arrival, our compatriot was taken with a serious illness. Thanks to the kindness of M. Tcheou I was able to send a wireless message asking our friend, Doctor Béchamp, head of the French Medical Mission and French Consul in Chengtu, for a consultation, although the wireless station had only just been installed and was not yet officially opened.

In the evenings I used often to go with Liotard to listen-in to the war news. We would return through the dark alleys, our hearts torn in anguish at the thought of the fearful disaster which was happening so far away, and from which fate had seen fit to remove us. We trembled at the thought that very soon the feeble ties which still bound us to our topsy-turvy world would be broken, and that the greatest events in history would run their pitiless course without our knowledge. We began to have qualms again about taking this plunge into the past, while our contemporaries remained to carry on the struggle. I had already wired to Dr. Béchamp[1] to ask him whether we should continue our journey. He never replied. He will never reply. He knows well that we can do nothing either for our country or for our people. We have been ordered to fulfil our mission, and we shall do so. But we shall carry with us a feeling of anguish which neither isolation nor the strange charm of our arduous profession will ever succeed in dispelling. Therefore, perhaps unconsciously, we delay our departure, in the vague hope that a miracle may still save our country.

Moreover, we are a little in advance of the lax time-table which we had arranged for ourselves. Although there are few indications of it—it snowed only the other day—it is still spring. And the best season for travelling in Tibet is autumn. In high summer the rainfalls are so frequent that they hinder the progress of the caravans. The soil of the upper valleys, drenched with water which the ice has settled on it, is dotted about with swamps and pools. The swollen streams are difficult to ford, and constant rainstorms fall on the travellers and soak their baggage.

[1] Dr. Bechamp died in a Vichy Government prison in Indo-China. He had been sentenced to twenty years penal servitude for not having despaired of his country, and for having joined with General de Gaulle after 1940. He was a man who astounded one by his truly encyclopædic knowledge. He led a retired life in his house in Chengtu, beloved by the Chinese whom he had learned to understand, and to whom he was bound by ties of friendship. Those few travellers who have been lucky enough to spend some days in Chengtu will understand that, having mentioned his name, I feel compelled to pay to his memory the modest tribute of my friendship and admiration.

The Himalayas cannot stop the progress of the clouds which the monsoon carries along with it.

In September, when the monsoon ceases to water the continent, the Tibetan sky shines again with its pellucid light. Then travelling is pleasant in the warm sunshine. During the day it is often more pleasant than in the heart of summer, for at that height the passage of a single cloud across the sun is sufficient to freshen up the temperature. Naturally the nights are very cold, but as the damp has gone, nobody minds. But from the end of November onwards the temperature sinks too low for the air to circulate and penetrate through the passes which, in Tibet, are usually sixteen to nineteen thousand feet in altitude. We hope to reach the districts which we intend to reconnoitre at the most favourable moment, that is to say in September, so that we may carry out our explorations in the best possible conditions. It will be soon enough, therefore, if we leave in the first week in June.

. . . . .

When we think of this departure now so close at hand, to which we have so long looked forward, a feeling of numbness comes over us: a feeling which the old archeologists in the provincial capitals of France, who have devoted forty years of their lives to studying the past, and who are suddenly wrenched back into their own period by the cruel reality of an air-raid, must be sharing with us. As I re-read the conditions of our mission, I can hardly realize that we are about to start off to determine the site of the sources of a river and to study how the folds of certain mountain ranges hinge into one another, while our native land is crumbling to pieces. We feel doubts, and shall continue to feel doubts about starting, up till the very moment when we take the plunge. And yet we shall do it. Since we no longer exist as far as the French community is concerned, since the ties which bind us to it are already nearly severed, we have no right to withdraw from the mission with which it has entrusted us, on the plea that fate has decreed against it. Whatever success may be granted to our mission, we shall be working for that French community, by adding to its spiritual heritage if we succeed, and by offering our exertions, perhaps even our lives, if we fail. Moreover, the task which we have voluntarily undertaken is not one to blush for, even in time of war. Liotard and I are fully aware of the dangers of our enterprise, and the information which we have gleaned from the various works written about Tibet have given us a sombre enough picture of the country of the Ngolo-Setas, which is our goal, to enable us on balance to banish our scruples.

The Ngolos! The first traveller to mention them was that delightful explorer who cradled our childhood with his admirable account of his journey across Tibet, Père Huc. He called them the Kolos—the spelling in no way alters their reputation—and he had never met them. He was content to record that there were "no monstrous practices which have not been ascribed to them". Nearly a century has passed since then, and many travellers have grazed the Ngolo country. All of them, less naïve perhaps in their language than Père Huc, have returned home with the conviction that it is one of the regions of our planet where, miraculously preserved from the assaults of time, the barbaric age of humanity has persisted. Thus, the French geographer Grenard actually wrote in the *Géographie Universelle*[1] that "this district is the least-known district in Asia and in the whole world".

For a territory to be able to preserve its mystery up to the present day there must be some important reason: either physical obstacles have made access virtually impossible or the peoples who inhabit it have guarded it against invasion. The last explanation fits the Ngolo country so well that an explorer, Georges de Roerich, concluded that the Ngolos composed no real ethnological group, but were merely a band of criminals, outlaws, who had taken refuge in this district and instituted a reign of terror. *A priori* I do not believe much in this explanation. That such a peculiar social phenomenon should have continued so long seems to me well-nigh impossible. The

[1] *Géographie Universelle*, Volume VIII (Haute-Asie, p. 355).

Ngolos and their close relations the Setas—so close that their names are often linked with a hyphen—must, on the contrary, compose a sufficiently homogeneous group to have preserved this independence up to the present day. What is certain is that the fear which they inspire has kept all explorers at a distance, and that we are to be the first to attempt to cross their lands.[1]

Technically, the object of our journey is to identify the sources of the Tong, one of the tributaries of the Blue River, and to demarcate the basin of the Blue River from that of the Yellow River in the Ngolo country. We intend at the same time to determine the direction of the orographic folds which divide the basins of the two large rivers, by locating the eastern limit of the Bayen-Khara mountains. Although the principal object of our expedition is geographical investigation, we are determined not to neglect the study of Man, and to devote ourselves as much as possible to ethnographical and anthropological observations, in spite of the danger to which experimentalists are exposed in a country like Tibet, where their devices might be mistaken for magic rites. But if people were overcautious they would achieve nothing at all, and would in the end shrink from bevelling away a rock with a geologist's hammer. Then caution would prevail to the extent of sitting quietly at home.

Before penetrating into the territory of the Ngolos we have to cross districts already known. According to our instructions we are to follow the valley of the Tong—which is known to the people round Tatsienlou as the Ta-kin, or the Ta-Tou-Ho—but on closer inspection it occurred to us that in so doing we should very soon run up against difficult gorges. Therefore, we decided to approach the Ngolo tableland northwards from the district of Luho, which is situated on the big commercial caravan road which connects Tatsienlou with Lhassa; the valleys there are less hollowed out by erosion and should offer less hindrance to the progress of the animals.

There are various ways of reaching Luho. We decide to

[1] For the confines of the territory of the Ngolo-Setas see "Geographical Notes" at the end of the volume.

follow first the trail leading from Tatsienlou to Litang, leaving it at the village of Tongolo, and from there to continue up the valley of the Lichou, which, from what we have heard, is unknown to geographers. It will be an opportunity for setting to work again and getting our hand in. Nevertheless we shall not be the first Westerners to visit this valley, nor the monastery of Lhagong which is situated at the source of the Lichou. A few years ago, Père Doublet, driven out of Tao Fou by Chinese Communists, took refuge there after some magnificent adventures which we often mention in our fireside talks. And then again, less than a year ago, the enterprising wife of a French diplomat Madame Georges-Picot, took it into her head to go there quite simply as a tourist. After Lhagong we shall rejoin the large pack-trail at Taining, where the important lamasery of Gata is situated, and travel along it as far as Tao Fou and Luho. We shall tread the path of our countless predecessors, missionaries and explorers, Frenchmen and foreigners. Only after leaving Luho shall we plunge into the unknown, into the territory of the Ngolos.

But before reaching Luho we shall pass more Catholic missions, those of Tao Fou and Charatong. These missions and the missions in the valley of the Salouen which we visited in 1936–37, are the most isolated missionary stations in the world—the real outposts of Christianity. Unfortunately, there will be no fellow-countrymen to greet us. The Charatong Mission is in the charge of Père Ly, a Chinese priest, while a few weeks ago, Père Doublet, of the Mission of Tao Fou, resigned his post, after a sojourn of fifteen years, in favour of a young Chinese missionary, Père Yang, who speaks and writes excellent French. On our journey to Luho, therefore, we shall be alternating between lamaseries and Catholic Missions. A strange country, Tibet, where a large part of the population is dedicated to the monastic life, and where foreigners penetrate only in order to preach their religion! But whatever their gospel may be, these Catholic Missions are havens of rest for French explorers, where they can prepare their minds and bodies for fresh trials. All explorers who have travelled through Tibet regard the missionaries somehow as companions-in-arms.

. . . . .

We spend the last days before our departure in taking lessons in the Tibetan language, in the library of the Mission House. Our teacher is a scholar who spent ten years in Lhassa. His good humour and natural politeness remove all stiffness from our studies. He has twice invited us to have buttered and salted tea with him *à la Tibetaine* in his little house, seated facing the domestic altar decorated with lama statuettes. Unfortunately, our Tibetan teacher is not the kind of Lhassa man to be of great service to us in the territories we are visiting, because the Eastern dialect he teaches us is in many ways different from that used in Central Tibet.

Our friend, Thou Ten, as he is called, also takes great trouble to find servants for us. Truth to tell, we are by no means overwhelmed with requests. In spite of the promise of wealth—there is a great deal to be gained in the service of Europeans—few are tempted by the prospect of a journey into the Ngolo territory. A few days after our arrival two strong, determined-looking men came to interview us but, after consulting a monastery oracle, they sent a polite message to say that the fates were against them, and that they were sorry not to be able to join us.

A rumour has got about that we are game-hunters, which increases the suspicions of possible candidates. It was Madame David-Neel's adopted son, Lama Yongden, who started the rumour. He is himself rather incensed against Liotard, who recently succumbed to the temptation of bringing down a vulture which had been attracted by the carcass of a horse killed descending the mountain. In Tibet one must be very cautious in one's treatment of animals. Shooting is always regarded with disfavour. Vultures, above all, are almost sacred by right of their dismal function of devouring human remains. "Killing a vulture is more or less the same thing as killing a man," Yongden used to repeat over and over again to us. We take his warning and promise never to shoot except when we are forced to do so to replenish our stock of food. Personally that does not worry me because I take no pleasure in shooting animals.

It was Yongden who finally managed to procure two servants for us, an uncle and nephew. The uncle, Yong Rine by name, is a Tibetan from Tatsienlou, and an experienced traveller. He looks a regular adventurer, with a low and entirely shaven skull, but it appears that his friends and relations are people of means and can vouch for him. Tchrachy, his nephew, is more prepossessing in appearance. I imagine he is a Chinese half-caste. He has had some education, and his polite manners and delicately modelled face put him above the ordinary caravaneer. He looks a trifle effeminate, for there is no trace of down upon his lip, but his courage has already been put to the test. He has done several dangerous journeys and has even been wounded by a stray bullet in some foray or other. We shall have to accept Tchrachy and Yong Rine because we simply have no choice! Their task will be not only to tend the animals of our caravan and to pitch our tents, but also to introduce us everywhere into the lamaseries and villages and to reassure the natives. From our point of view, therefore, they are of the greatest importance. This they thoroughly understand and, proud at the idea of serving men who are important through their connection with the Marshall and also because of their financial means, they pose everywhere as the confidential agents of exalted personages.

In addition to these two Tibetans our personnel consists of a Chinese cook, whom we engaged in Chengtu to go with us as far as Tatsienlou. Good old Tze! He fell in so well with our habits and we with his cooking that when he asked to be allowed to continue with us on our journey, I could not refuse. I fear, however, that the good fellow, no longer young, does not fully realize what kind of expedition he is in for. I have a vague feeling of remorse at dragging along this silent and reserved peasant from the Sseu-tchouen plains on our adventurous journey across the tablelands of Tibet.

. . . . .

Louis Liotard.

Tatsienlou.

A saddled yak.

Our two carriers; Tchrachy on the left and Yong Rine on the right.

Tibetan landscape.

Near the monastery of Lhagong.

The lamasery of Lhagong. On the left is a chortain*—a typical Tibetan monument.*

The lamasery of Lhagong.

A costume and mask worn by the lamas in their ritualistic dances.

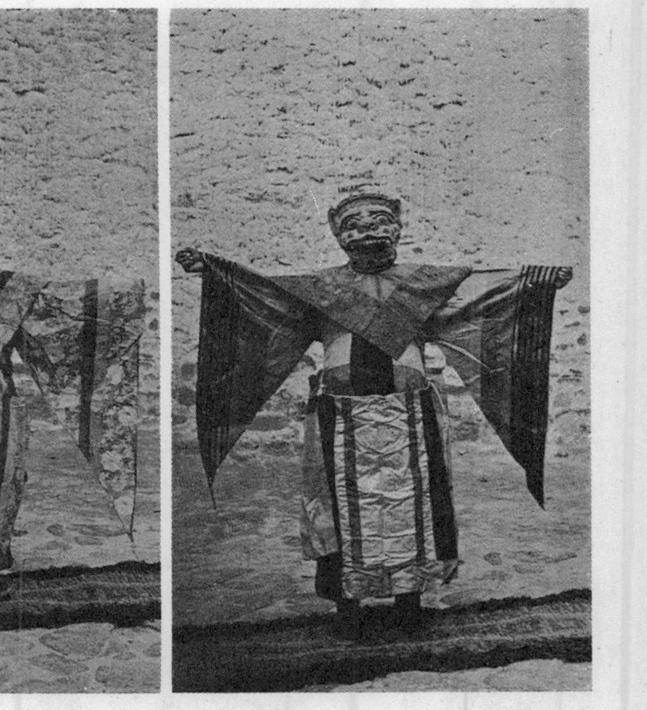

Other examples of masks and costumes worn by the lamas on ceremonial occasions.

Dignitaries of the lamaistic cult officiating before an open-air altar. The grand lama wears spectacles.

A ceremony between ritualistic dances. The idol placed under a Khata takes the place of a human sacrifice.

In the annual procession of books from a monastery library a palanquin is carried by the monks containing a statue of divinity.

In the procession monks and novices each carry a volume.

The valley of the Tao.

The valley of the Tao.

Chortaintong. Near the site where the brigands attacked the Guibaut-Liotard expedition.

An example of Tibetan architecture.

A Tibetan with his gun which is provided with a double-pronged rest hinging about half-way along the barrel.

Shepherd boys of the Tibetan highlands.

Tibetan peasants.

Peasants in the courtyard of the Tao lamasery.

A lama holding his rosary.

Earthquake Shooting

Last night there was an earthquake. To tell the truth I did not realize it till this morning. Woken with a start by tremors which made the badly-fitting glass door panes tinkle and imagining there was a thief in the house I got up, revolver in hand, and jerked open the door. My surprise and alarm were great when I found no one in the courtyard. This morning Monseigneur explained the riddle to me.

We are in a zone of considerable seismic activity. In 1934 there was a terrific earthquake to the west of Tatsienlou. Père Doublet tells us that we shall see traces of it everywhere, because the earth has not yet healed up. The rest of the world has known nothing of this upheaval of a whole district. Thus, in this vast continent of Asia, disasters can occur without anyone elsewhere knowing anything about it.

In Tatsienlou even minor earth-tremors are a constant source of danger. The houses are so huddled together at the foot of the steep mountain slopes that they are always in danger of being crushed by loosened rocks. Such accidents occur sometimes even when there is no earthquake at all.

My night's misadventure is the cause of friendly raillery on the part of our hosts, the missionaries. It is an opportunity for them to recall the funny incident which occurred in Bahang, on the Tibetan frontier of Yunnan, when a Père seized his gun and rushed out of doors to shoot an earthquake! The word which in the local dialect means "earthquake" is phonetically so similar to the word for "tiger" that, startled by the cries of his servants, whose sharper senses had detected tremors, our compatriot imagined that one of these felines was in the neighbourhood. One can well imagine the astonishment of his entourage, wondering what on earth the Père was going to accomplish with his rifle.

This impulsive missionary may have had small hope of finding a tiger at his door, but on the other hand nothing would be more banal than to discover a thief in the courtyard of the Mission House. Robbers abound in this frontier town, making frequent night incursions into gardens and private houses. It is not good to be found in the streets after ten o'clock at night. It is quite common at dawn to find people lying stabbed and entirely stripped of their clothes. Only the gates of the town are guarded by soldiers; in the interior the rounds are performed by night-watchmen. As these rounds are done at fixed hours, and as these casual guardians of the peace take care to announce their arrival with rhythmical vibrations of their gongs, it is difficult to see how they could hope to surprise a thief. The inhabitants of Tatsienlou are not greatly concerned about these dramas of night life, and merely take the precaution of not going out in the evening.

But the other day they were frightened in quite another manner when an aeroplane circled over the town. Everyone anticipated a bombardment. And yet the distance which separates us from the front line makes such an attempt unlikely although, from another point of view, such an idea might easily take root in the minds of those ridiculous Japanese who, in this pleasant month of June, are busily engaged in dropping tons of bombs on Chungking and other towns of the Sseu-tchouan without much valid military pretext.

For some peculiar reason the Tibetans seem less impressed by the aeroplane than those Chinese inhabitants of Tatsienlou who have never seen one before. Steeped in the marvellous and the uncanny by the rites and practices of their religion, they bow more readily to the unaccountable than the Chinese realists. When one has admitted once for all the existence of the supernatural, there is no longer any reason to be surprised at anything.

. . . . .

To-night, *June* 12, is the eve of our departure. To-morrow we shall be leaving Tatsienlou; once more we shall be bidding farewell to fellow-countrymen who, as always happens in remote places, have become our friends. The memory of them will follow us, and we shall talk of them in our nights of depression or in our more expansive moods. And to the melancholy of these last hours, which we cloak under a make-believe of frivolity, is added a new and unexpected sorrow, not the tightening of the heartstrings at the possibility of our never returning, but the frightful certainty of the collapse of France. M. Tcheou has just arrived bringing the news that Italy has declared war against us. That is the death-blow! So we are going to take the road with that wound in our side.

We stay up late beside the burning brazier talking to Monseigneur and Père Doublet. In the hazy atmosphere caused by the smoke-rings from our pipes, we all try to keep up our spirits, the missionaries to give us courage, we in order to divert their minds. Outside the rain pours steadily and monotonously down, drenching the courtyard where our fourteen horses and mules trample about in the thick mud. We are cold, sitting in our woollies beside the charcoal burner, and the damp is so penetrating that the paper in our note-books reacts towards ink like blotting-paper. And yet we are only at an altitude of 9,500 feet. To-morrow we begin our ascent towards the real Tibetan tablelands, and never again until our return into China shall we be at such a low level as we are now.

Our Tibetan servants have gone off to spend their last evening in town. Tze, the Chinaman, has gone quietly to bed. Every now and then the voice of a muleteer rises above the tramplings and buckings of the fractious animals. This caravan does not belong to us. We have hired it to take us as far as Tao Fou, where we intend to buy the animals which will form our real caravan. So our personnel is increased by three mule-drivers, whose business it is to return the animals to their owner in Tatsienlou. The fellow was very reluctant to entrust them to us because the roads are by no means safe. The journey out will be safe enough—there are several of us, all armed. But the owner is frightened of the return journey with only three men in charge of a caravan which might at any moment be attacked by looters. Here, as everywhere in Tibet, the only danger is from brigands. There are numbers of them on the Tao Fou road and they have no qualms about attacking. Their principal victims are travellers returning from small gold-washing expeditions; they fall upon these poor wights and deprive them of the few grammes of gold dust which constitute their whole fortune.

We, for our part, are about to file past these looters with what, for this country, is the equivalent of a fortune. We have to take with us enough money not only to keep ourselves alive for at least a year, but also to buy a caravan. Chinese banknotes are not legal tender in Tibet, so we have swopped them for silver coins, some of which are only silvered on the outside and stamped with the image of some age-old emperor. These rupees comprise one entire mule-load. To lessen the risk we have divided them up and put several in each of our various cash-boxes so that if one mule-load disappears or falls into the river the disaster will be less overwhelming. With our scientific implements, our photographic and cinema apparatus and our camp paraphernalia, we shall be quite a little treasure-party joy-riding along the mountain tracks of Tibet, exposed to all the hazards of the road. Everyone will hear about us because it is impossible to keep our expedition a secret. I look forward to the prospect with certain misgivings.

. . . . .

We take leave of Père Doublet on a little stone bridge which straddles a stream about a mile beyond the gates of the town. Tears stand in our eyes and we clasp hands with deep feeling, exchanging the customary little jokes. Then we get quickly into the saddle while our friend returns to Tatsienlou which, we realize only too well, will soon live in our imaginations with all the charm and advantage of "town life". We know that, modest and unpretentious as it is, we shall find nothing on our journey in any way comparable to the city we have left behind.

Mounted on our horses, magnificently caparisoned with light-coloured saddle-cloths, we take our places at the head of the caravan, and the journey starts.

Yong Rine and Tchrachy look fine seated on their mules with their daggers, their broadswords inset with precious stones, and our magazine rifles slung over their shoulders. On their heads they wear fur caps, and their necks, ears and fingers are adorned with silver rings. Their large reliquaries jolt against the left side of their bodies, and their right sleeves, which in the Tibetan fashion they wear loose, hang elegantly down against the right flanks of their horses. The three mule-drivers on foot, dressed like comic-opera gipsies, urge on the pack-horses by throwing stones at them and exchange coarse jokes with our hirelings.

In this little group Tze is the embodiment of the Chinaman, meek and compliant. He is dressed in the national blue and seems to have thickened out of all proportion. Under his coat he has put on every stitch of clothing he possesses.

Liotard's horse is almost a dwarf. Mine, on the contrary, is enormous. We must look something like Don Quixote and Sancho Panza.

The unknown adventure lies before us at the far end of this watery, paved road which leads through the forest up the mountainside towards the Tcheto pass, gateway into Tibet. Already we feel its stirring call.

CHAPTER 2

THE GREAT SANCTUARIES

IT took us seven days to reach the lamasery of Lhagong because our course was slowed down by the rain. I never imagined that the wet season would set in so exactly to time. It rained almost half the period of our journey and the ground was so dotted about with swamps that our feet and legs were constantly drenched by the water thrown up by the horses' hoofs. The rivers were swollen so high that it was dangerous to attempt to ford them.

Nevertheless it was in marvellous sunshine that we made our entry into Tibet, two days after leaving Tatsienlou. At an altitude of 14,000 feet the reflection of the sun on the snowfields of the pass was unbearable, even through my black spectacles. Lovely flowers, heralding the approach of summer, were already springing up through the carpet of snow. The dog of the caravan, tired after the climb, refreshed herself by rolling in ecstasy over the stainless white sheet and greedily gobbling down the white crystals.

For several days caravans had been waiting at various points for this rift in the clouds before crossing the Tcheto pass. Now they all rushed together towards the pass, blocking it with their yaks, mules and horses, while their riders jostled one against the other, exchanging a cheerful hullabaloo of jokes and oaths. And all the while vultures circled round high up in the sky, watching for the death of one of these creatures, man or beast, so that they might swoop down and tear it to pieces, leaving no trace but a bleached skeleton such as those we see strewn about us on the ground.

In the midst of this picturesque crowd, ploughing through it like a speed-boat through a fleet of fishing-vessels, appears unexpectedly the caravan of the Grand Lama of the monastery of Gata. It outstrips us just as we are reaching the ridge. The prelate has scorned to make the ascent on horseback. Hoisted on the shoulders of eight strong carriers, he remains inside his chair entirely enclosed by black curtains which resembles an enormous bier, only emerging for a moment to throw, with a pontifical gesture, a stone plucked from the ground upon the enormous pile of small stones and rocks which pious travellers have erected bit by bit on the pass. Then one of his dignitaries, dressed in yellow, with a gilt tiara on his head, dismounts from his wonderfully-caparisoned horse to burn incense and to place on top of the pile a prayer-banner, which nature's inclemencies will slowly transform into a piece of shredded cloth similar to those which now lie flapping in the breeze, left there by other caravans. . . .

We, for our part, from the top of the pass, see Tibet lying before us. The landscape is suddenly transformed. The outline of the slopes reaches out in harmonious curves, unbroken by any sharp ridges or crude formations, and sinks gradually downwards into the depths of the wide valleys; peaceful undulations of an ancient country moulded by the passing of the years. And, entering this district for the first time and recognizing it without ever having seen it (our studies having made us so familiar with its character), we shout aloud with joy. Below the snowfields the grass of the pasture-lands and of the whole landscape is withered by the cold into a uniform brown colour. Against this chestnut-coloured carpet the herdsmen's tents look like spiders, and the cattle like worms which have eaten into it. One can understand how the ancients believed that Tibet sheltered among its mountains fantastic animals unknown in any other country.

When we reach the bottom of the valley we meet our first herdsman. He is a handsome youth with a copper-coloured skin, his sheep-skin coat rolled down to his waist, preferring to expose his torso because it was so warm. The thermometer stood at 41 degrees Fahrenheit.

. . . . .

In spite of the rain our journey from Tcheto to Lhagong was delightful, just like a camping expedition. The valleys along which we travelled were wide and fertile, studded here and there with prosperous villages or single houses, topped occasionally by dzongs, those typically Tibetan castles, large and imposing, lodged firmly on the shoulders of the hills, flanked often by high, half-ruined watch-towers. These watch-towers, relics of a former age, and perhaps of an ancient civilization, have a star-shaped outline which astounds the traveller by the intricacy of its design. Beside the villages were well-cultivated fields of barley, surrounded by low walls, with here and there clusters of birch-trees and willows. We met several herds of yaks, sheep and horses browsing at liberty along the mountain slopes. A truly happy, gentle landscape, which summer was already beginning to adorn. Buttercups were already blooming in the grasslands, and violets and forget-me-nots gave the finishing postcard touch to this otherwise bleakish prospect.

When it was fine, camping was a delight. You only need dip a line into the river to find a trout hanging on the end. This was the origin of an odd incident which occurred close to the little monastery of Posang.

We had already caught three fish and were preparing to bait our hook once more when three horsemen came riding towards us at the gallop. With many apologies but also with insistence they told us that we had caught enough fish for our meal and that there was therefore no object in our continuing to dip our lines. Fish are held sacred in Tibet, and this doctrine has given all Tibetans, even nonconformists, an extraordinary distaste for this dish. Apparently in some Chinese-inhabited districts the more cunning of these take to fishing, selling their catches to pious Tibetans who buy them in order to acquire virtue by returning them to the water. Nothing could more aptly describe the difference in mentality of the two races.

We met with no serious difficulties up till the last stage of the journey. The valley had become a narrow gorge, the river a torrent, and our mules had to force a passage through a jumble of granite rocks. Every now and then we had to unload one of the animals to enable it to clear the obstacles.

Finally, at a height of 13,100 feet, we reached the plateau where the Lichou takes its source, a vast brown plain fringed with mountains, in the centre of which stands the monastery of Lhagong, the only human construction in this vast panorama. The Lichou was nothing but a broad, lazily-meandering stream which traced wanton patterns across the marshy plain. In past ages it had apparently changed its bed at will, because there were banks everywhere indicating where it had previously flowed. Yet even up here it was fairly deep, and our horses in crossing it sank breast-high into the flood. Not a single tree, not even a shrub was to be seen in the vast expanse which lay before us. We had come into another world; we had reached the roof of our own world.

With proper decorum, in the ritualistic direction, that is to say clockwise, we made a complete circle with our caravan round the monastery before halting in front of the big gate. In this manner the lamas, who were disconcerted by our sudden appearance, were given to understand that we respected their customs.

. . . . .

We have made the customary exchange of visits with the "Lama chimbo", the Great Lama. He came into our room this afternoon, accompanied by his steward, to offer us a khata and a bottle of milk. He is a fine-looking man of fifty-three, tall in stature, rather dirty but with pleasing manners.

His monastery, which is of the sect of Satias, is a typical pastoral temple. It juts up isolated in the plain, without any attendant village, and its devotees inhabit the hundreds of scattered tents which dot the lazy swell of the hills like black-headed pins.

Facing the gate the main chortain rises white and bulbed. Above the house-tops jut the horns of the superimposed roofs of the great temple, with their little bronze bells tinkling in the breeze, and their beribboned poles topped by copper balls outlined against the sky. Under a wooden penthouse or cloister are a hundred or more prayer-mills cogged one against the other, with the same monotonous inscription painted in large Tibetan lettering on their cloth sails: "*Om mani padme houm, om mani padme houm* . . ." In this cloister there is a constant coming and going of lamas and faithful who keep mechanically turning the wheels, which are of varying dimension and design, and then circle round the large chortain before finally entering the monastery.

The doors into the various sanctuaries are at the back of the inner court. They are covered over with wide black draperies edged with white like the draperies used at public funerals.[1] Prayer-banners also cover them, hung there by pilgrims; these prayer-banners are gradually transformed by the passage of time into filthy colourless rags.

The court and its façades may give an impression of dirt and decrepitude, of some inspiration for ever unrealized, but by contrast the interiors of the sanctuaries are quite clean. On an altar which stretches the whole length of the temple, one sees, rising up in relief against the rear wall, in the pale light of the butter lamps, huge statues of gods and goddesses, richly painted and dressed in silk and brocade like the Spanish madonnas. The architect has not troubled to arrange them to advantage. One god, for instance, twelve feet in height, is imprisoned in a crevice hardly six feet wide. You have to approach right up to his feet and gaze up at his fore-shortened figure from below.

Softened by the half-light, the multi-colourings of these temples shed a warm, comfortable glow. Every inch of space

[1] It should be explained that in Tibet, as in China, black is not the colour of mourning. In China white is the colour which symbolizes death.

is decorated; the rafters, the wooden pillars, the doors and even the shelves along which are ranged the articles of worship. Surrounded on all sides by these warm lively tints which clash one with the other only to melt again into a harmonious whole, we begin to regret the soberness of our own tastes and our tendency to monotony. The riot of colour reaches its climax on the cloth-draped walls ornamented with frescoes.

Strange paintings these, in which neither people, animals, plants nor landscape are native to Tibet! And yet the artists who painted them have never left these districts. I have actually seen them at work here, in a cold, gloomy attic overlooking the monastery courtyard. Through the centuries they have continued their tireless task of painting Hindu personages or Chinese flowers. Apparently it has never occurred to them to copy the surrounding landscape. But what they have added of specifically Tibetan to these inspirations from abroad suffices to give their paintings a rather morbid flavour and originality. It is a kind of mixture of the sadistic and the macabre, which testifies to a sublety and tortuosity of mind astounding in people who lead such simple lives. Some of these paintings are portrayals of lust: frightful deities riding rough-shod over the naked bodies of men and women, whose agonized attitudes seem intended to inspire sensuality rather than terror. Occasionally one sees the act of copulation nakedly portrayed, with gods twining their many arms cruelly and passionately around their *Çakti*, their female counterparts, whose outlines are lengthened lewdly and out of all proportion, their bodies bent backwards, their legs folded under their straddled thighs, their waists gripped by clenched hands with pointed nails which dig deep into their flesh.

It must be admitted that the enthroned Buddhas squatting Indian-fashion on lotus blossoms on the main altar have no such horrific appearance, and seem to float serenely above the skies in which the genii and demons, who hold a less important rank in the absurd Lama pantheon, roll around in the grip of their passions and follies. Their majestic calm, their Olympian serenity harmonizes with the peaceful landscape which surrounds them, and they have even something physically in common with this landscape, whose well co-ordinated curves seem to obey the law of natural phenomena, which tends always towards balance and stability.

As I come out of the monastery I cannot help noticing with surprise the tents of the herdsmen encamped around. So these wretched people have amassed fortunes to build, on this plateau, this gorgeous sanctuary, which harbours hundreds of works of art, without ever thinking of building houses for themselves. It does not appear to worry them and they seem happy to be privileged to spend a few days in the shelter of this monument, which is the centre of their little universe. It is the hour when the herds assemble, and when the plain becomes more lively: the young men start races on horseback, and riders return at the gallop, with their forked guns slung across their backs, swaggering a little as horsemen always do.

Then, with the twilight, sounds the loud, sad call of the copper trumpets of Lamaism, great horns which extend to nine feet in length, and a lama comes and seats himself at the foot of the great chortain. A few of the faithful bow down before him, their heads to the ground. He touches them on the forehead and then offers them a little sour milk in a spoon. I stop in front of him. He hesitates a moment and then holds out to me the ritual offering. I bend down and put my lips to it. The good man appears enchanted.

. . . . .

Between Lhagong and the monastery of Gata we were lucky enough to see simultaneously two of the highest mountain peaks in the world, Mynia Gonka and Jara. The first, which is the highest point in the whole Chinese Empire, is 24,900 feet high. It has been scaled by two German mountaineers. We could hardly distinguish it from our own altitude of 13,100 feet for it was about sixty miles away as the crow flies and, in any case, relatively speaking, it was from our point of view only a small mountain of 11,500 feet. Jara, which lay much nearer to us, was more impressive. Its peak rose up perfectly cone-shaped, and the dazzling whiteness of its glaciers cast an even darker shadow over the brown hills and plainland. Although other travellers have been more moderate in their estimate, I myself believe Mount Jara to be more than 19,000 feet high. In any case, the regularity of its outline and, above all, its isolated situation make it one of the most beautiful mountains of the globe, and it might well tempt all those who like to be "the first". Looking at it from the west, I should imagine that no one could make the ascent alone.

We have now been two more days on the march. Our course took us first across a prairie dotted all over with herds of cattle, and subsequently, after crossing a small ridge, right down to the bottom of broken valleys, hollowed out by erosion or sliced into the tableland by gigantic landslides, remains of geological upheavals. We got down as low as 11,800 feet, and then immediately rose again, continuing the everlasting "scenic railway" route of Tibetan travellers. We passed quite close to the lamasery of Pamé, perched on a high eminence and protected so much by escarpments as to be apparently unapproachable. We did not stop there. And from now on we shall see monasteries dotted about everywhere, planted in sites specially chosen for their ability to resist a siege, all of them similar in their serene majestic isolation in these stern natural surroundings, and yet all radically different in design and construction. And in spite of our longing to enter them, in spite of the appeal which these havens of rest already have for us, we cannot stop at every one of them. So we give a wide berth to these sacred islets, whose incense fumes reach as far as our caravan, as the scent of Italian islands reaches the sea-bound ships, imbuing the sailors with a yearning for countries and peoples which they will never know.

Nevertheless, after camping yesterday in a charming little deserted valley, we are at present the guests of the important lamasery of Gata. We suddenly caught sight of it in an amphitheatre of hills, in the centre of a plateau; a great walled enclosure with temples and many private lama dwellings. Its horn-shaped roofs vie in stature with the tall poplars which spring up everywhere between the buildings, although there is no tree of any kind on the plain. Like the pagodas in China, the Tibetan sanctuaries thrive in the shadow of old stately trees.

Gata Gomba is one of the great lamaseries of Eastern Tibet. In spite of its isolation and its outward appearance of calm it has known troubled times since its foundation in the early eighteenth century. An earthquake overthrew it in 1893. In 1905 it was pillaged during a war between China and Tibet. Then again it was plundered in 1935 by a passing band of Chinese Communists, but rebuilt again shortly afterwards. After Gata Gomba the houses of the Chinese village of Taining, founded upon the mud of this spongy tableland, appear more than usually squalid. The Tibetans build better than the Chinese.

This time our hosts are the Gelukpas, the reformed sect who worship the Dalaï Lama, the first divine reincarnation of Lamaism. Tsong Kapa, who in the fifteenth century started the reform, the object of which was to protect the purity of the religion, has himself become a divinity. He thrones it on the high altar of the great temple, a colossal stucco effigy with placid features, dressed like a lama, with a large yellow cap on his head, squatting Indian-fashion on his heels on a lotus flower in full bloom.

The lamas who conduct our tour spare us no details, and obligingly satisfy our curiosity on all points, heedless of the sacrilege of our presence in these holy places. And their attitude is not inspired by the desire for gain—they would be extremely surprised if we offered them a tip. Contrary to the opinion generally held in Europe, Tibet, as indeed the whole of the Far East, is a country of tolerance. Here, as in China, the question "To what religion do you belong?" would be quite irrelevant. The Tibetans have so many gods that they are quite ready to acknowledge that foreigners should have others, and are willing to respect their existence. Thus they are often on the best of terms with ministers of foreign religions. Père Doublet told me that when, after a stay of fifteen years in Tao

Fou, he informed his neighbours, the lamas, that he was due to leave, the latter suggested that they should address a petition to the bishop asking that he should not be recalled. One cannot be more brotherly than that.

We are allowed to go anywhere, to walk freely through the temples, to touch the religious symbols, to take films and photographs, and no one raises the slightest objection. On the contrary, everyone does his best to help us, and takes an intelligent interest in our procedures. Never in any country in the world have I seen so many smiling faces. This natural friendliness in the Tibetans goes to the hearts of travellers and partly explains their attachment to the country.

Our jolly guides make laughing comparisons between the appearance of Maïtreya, the future Buddha, saviour of the world to come, and our European faces. This bodhisatva, who, according to tradition, will begin his work of salvation by opening the tomb which contains the still fresh body of Kachyapa, forerunner of the Buddha Çakya-mouni, is depicted seated European-fashion on a throne, his legs dangling, his face and body lightly painted. They say that he will manifest himself as a white man. Are Liotard or myself going to be taken for the future Buddha?

This sane and reassuring effigy restores our peace of mind after the horrific aspect of the gods who haunt the neighbouring sanctuary, a little tantric temple set back from the great temple, which is merely the nightmarish expression of a fevered imagination. In every monastery there lies hidden away a place reserved for the mysteries of Tantrism, that strange doctrine, or rather ensemble of monstrous practices, which has passed from the Hindu cult of Çiva to Lamaism. Everything that the extravagance of oriental wizardry would possibly conceive, everything that the disordered minds of hermits could invent is to be found in this luxuriant ritualistic display, in which cruelty and eroticism are interwoven. Stuffed animals, a pheasant, a small brown bear and even a cow hang down from the ceiling, sorry-looking creatures, putrefying, with the stuffing hanging out through the holes in their bellies. The bear's eyes have been removed and replaced by conical-shaped pieces of wood with eyes painted at the tips. These stalked eyes evidently play an important role in the tantric creed, for on the door of the temple there is a frightful-looking bearded mask, with staring eyes projecting at the end of stalks. The interior of the temple is filled with monstrous scenes of copulation symbolizing the unification of masculine and feminine power, which, according to this doctrine, is the expression of supreme wisdom.

The God who rules over this world of lust and folly is indeed worthy to occupy the chief place on the altar. He is smeared all over with blue paint, has three eyes and wears a crown and necklace of human skulls. In his eighteen arms he brandishes the most extraordinary assortment of objects, a thunderbolt, goblets carved from human skulls, an elephant, a small bell, and many things besides, and against his body draped in a tiger-skin he strains the entirely nude figure of his female counterpart. The two are depicted in wild demonic frenzy trampling upon a row of shapeless gasping corpses. I do not believe that anywhere else the human imagination has run riot to such a point. As though the guardians of these maniac phantasies were vaguely aware of their significance, they have spread a light muslin covering over these monstrosities, which recall an age, perhaps not so very distant, when the performance of these rites was accompanied by human sacrifices.

Ancient weapons hang from the pillars and walls of the temple, old flint guns, sabres, lances, bows and leather quivers, such as the barbarians used in their former invasions of Europe. There is even, on a wooden stand, an iron suit of armour, obviously very ancient. Why this armoury? Is it to prove the savagery of man in relation to that of the gods whom they have created for themselves? It would be just as difficult to decide this point as to determine why the frescoes here are toned down into red and white designs, standing out in relief against a black background.

. . . . .

The ruler of the Gata monastery is none other than the Great Lama who outstripped us on the Tcheto pass. He welcomed us very kindly, reminding us of our meeting and of the civilities which we exchanged on the mountain summit. He is a man of considerable political importance, being a member of the Committee of the Si-kang province, a kind of Council Chamber which holds periodic sessions at Tatsienlou. We found him in the temple, busily supervising the arrangements for a religious ceremony, and he at once invited us to remain another day in order to witness it. We accepted all the more readily because the weather was still abominable. The snow has sunk very low on the mountain slopes, and since it does not hold at the height on which the monastery stands, there is a constant fall of icy water over Gata.

We have the good fortune to be comfortably housed. A Chinaman, M. Lieou, a Gold Trade inspector, has put at our disposal a room in the lama's house just beside the temple, which he occupies with his wife and child. The ritualistic murmurings disturb our leisure hours. Our room is entirely carpeted with Chinese newspapers, and instead of window-panes we have squares of paper. But it is less cold than in the tent. Already, after ten days of travelling, this sordid room seems like a palace, and we relax our tired bodies, thankful to be sheltered from the inclemencies of nature.

. . . . .

The service, like church services all over the world, starts with a hymn sung in chorus. It is an intoned chant, which resembles our plain-song, swift and rhythmical, but more monotonous and with a duller resonance. The low notes, coming from the depth of the throat, hummed as in certain Russian folk songs, sound hollow like the notes of a double-bass, causing as it were tactile vibrations in the air.

There are at least two hundred monks and a hundred peasants in the temple, all squatting Indian-fashion along long strips of red wool. Instead of facing the altars which align the far wall, the congregation is formed into two opposing groups each backed against the side walls. Thus the statues of the gods, standing erect in the half-light over which their lively colourings cast a faint glow, appear to be looking down in sublime detachment upon this human gathering.

In the front row of the left-hand group, the Grand Lama, decked out in dazzling canonicals very similar to those of a European prelate, presides over the ceremony. Only his yellow felt tiara, shaped like a hoplite's cap, which lies on the pulpit in front of him, would strike a discordant note under the arches of a cathedral. He himself is not squatting, but is seated on a raised chair adorned with rich brocades. To his right, at a slightly lower level, sits the prior of the monastery, and opposite him the young living Buddha, draped in a violet toga. This young god of flesh and blood, frail and anæmic, squats on a low table, like a deity on a lotus.

In between the rows, the lama masters of ceremonies pass up and down with censers. Blowing down long pipes they puff under the noses of the assembled crowd great clouds of smoke which rise upwards in flakes in the still ether. The air is heavy with mingled scents, the perfume of the incense, the steam from the buttered tea and various human exhalations.

The faint light falling obliquely from the upper galleries which support the only windows of the sanctuary casts its rays over the customary Tibetan colour symphonies: the faded red of the monastic robes, the bright yellow waistcoats, the purple of the lay tchoubas, the reddish brown of the necks, faces, shoulders and bared arms, and the bronze tints of the shaven skulls.

While the congregation continues its murmuring chant, the ceremony begins. A procession of monks in single file lays offerings on the pulpit of the Grand Lama. These offerings look a bit out of place, large tsampa cakes ornamented with coloured butter. It is like being in a pastry shop. The Grand Lama performs over these set-pieces certain professional ritualistic gestures, and then hands them over to a lay assistant who immediately places them on a table. He is obviously in a hurry to bring these preliminaries to an end.

The saintly man, having dispatched the last cake, is immediately presented with large, white scarves, and the pastry-cook is transformed into a draper measuring lengths of material. With quick nimble gestures he spreads them over the enormous book which lies in front of him. He only resumes his priestly unction when, replacing his peaked cap upon his head, he lays his muslin-covered hands on the heads of the five officiating monks, who bow reverently before him. This gesture is a signal for the congregation to release a whirl of khatas, which fall gently and noiselessly at the Grand Lama's feet like a flight of doves. For the remainder of the ceremony they will lie there, between his chair and the table of the living Buddha, in a thick snow-white heap. This charming performance is the last act of ritual before the commencement of the sermon.

I find this sermon a little tedious for I do not understand a word of it. Nevertheless I admire the speaker's eloquence. The holy man addresses his audience simply, one might almost say familiarly. It is clear that he does not mind mixing jokes with religious instruction because his listeners laugh from time to time. The attention with which these good people follow his flow of language is really admirable. The young monks, aged between eight and nine, stop giggling, and the peasant women leave off picking their lice.

The sermon ends with a general distribution of hot buttered tea. Everyone produces from his clothes the wooden cup which he carries about with him everywhere and holds it out towards the masters of ceremonies who go up and down with steaming kettles.

The Grand Lama is truly indefatigable. Having poured one bowl of tea after another down his throat, he immediately enters upon a fresh task. I can well understand why children destined for a religious life have to undergo a severe preliminary medical test. The practice of religious rites sometimes demands real physical endurance.

Opening the large volume which lies in front of him the minister starts reading aloud at an incredible speed. He is such a master of the art of drawing breath at the right moment that his delivery seems to flow uninterruptedly. Is he going on like this to the very last page of the book? I am beginning to fear so. Then, to my relief, I perceive that he is only making a pretence of reading. Only the first few lines of each page are uttered in an intelligible manner; the rest of the text is lost in a sort of humming which is incomprehensible even to the most practised lamas. From time to time this low intoning changes abruptly into some ritualistic formula, taken up in chorus by the congregation, or into the inevitable "*Om mani padme houm*".

And so we go on, one page after another! As it is impossible to follow the reading the attention of the faithful relaxes a little. The novices start fidgeting and whispering to each other things which probably have nothing to do with the ceremony. Others open their mouths wide and start yawning. Even those lamas who cannot plead youth as an excuse disregard the ritual to the extent of talking to each other, and the living Buddha himself leans twice against the shoulder of a hoary old lama to confide something in his ear.

Tired of following with my eyes the flutterings of the tame sparrows in the temple, I let myself sink into a torpor. Hours have passed since we first entered the sanctuary. The light grows faint, and no doubt the sun has disappeared behind the mountain. It is suddenly cold, and many of the lamas, with the traditional shuddering gestures, have already lifted the flap ends of their imperial togas over their bronzed shoulders.

"*Om mani padme houm.*"

It is extraordinary how this continued mumbling of formulas can swamp all individuality. Huddled in my fur-lined coat I squat on my strip of wool, dozing and dreaming with the rest. I have even forgotten the presence of Liotard at my side. Through my half-closed eyes I see the Grand Lama like a pale gold dot in the distance.

"*Om mani padme houm.*"

This time I actually joined in with the rest in the mystic invocation of Tibet, the chant which is repeated over and over again, not only by the monks in the monasteries, but also by the hermits in their ice-bound retreats, by travellers on windswept passes, upon the praying-mills, upon the engraved stones of the mani and on the high peaks where one can read, in gigantic lettering, this mysterious inscription, whose meaning has been lost in the mist of ages. It is doubtless an evocation to the unseen world, the territory of unknown, uncanny faces. Where, better than here, on these high tablelands towering to the skies, can one recapture the terrors of the early ages of mankind?

"*Om mani padme houm.*"

Hauntingly the chant goes on.

Perhaps I am dreaming. I feel through these high stone walls the magic of this enormous desolate country which, by the thinness of its air, seems to be the roof of the world; the magic of this scorched land in which the mineral carcass of the earth springs up everywhere, thrusting aside the thin skin of life, and where man, that accidental creation, clings on by a thousand-year-old process of acclimatization, in such unsheltered conditions that he can only resort to prayer.

"*Om mani padme houm.*"

The wind blows gustily down from the glaciers of the Jara and makes me shiver.

The Jara! Lifeless mass of rocks, or pillar of this country—altar whose scattered acolytes, far removed from the plains where swarm the ant-heaps of humanity, cannot be like other men who breathe the heavier air of the lowlands! Nowhere but here, in this atmosphere, could the lofty conception of Buddha unite with the dark, primitive rites of ancient Shamaism, to culminate in the monstrosity of Lamaism.

"*Om mani padme houm, om mani padme houm.*"

Barriers between the visible and the invisible, between the real and the unreal. . . . These people, mindful of things which we cannot see, and which they themselves never perceive, are naturally disposed to find a supernatural explanation for everything. The sparrows chirruping in the temple are not sparrows to them. And why should they be sparrows after all? And the flames from the butter lamps which flicker in the rear wall of the sanctuary—are they really flames? Does anyone really know what a flame is?

"*Om mani padme houm.*"

We were awakened from our trance by the bowings of our neighbours and by the hubbub of the departing congregation. I felt a little embarrassed at being unable to follow the ritual and do as the others did. But perhaps I did make some sort of inclination before these great golden gods, draped in muslin. In any case I bowed low before the Grand Lama, who retired in great state, preceded by trumpets, seated under a canopy, escorted by other lamas swinging censers, thankful that the whole business was over.

Now we are walking down the muddy alleys of the monastery. The little monks scamper along ahead of us, glad to be able to stretch their limbs at last. We are not yet quite awake, and neither of us feels inclined to talk. As we reached the door of our house night was already falling.

"Paris finished! Taken! Paris is taken!"

I don't quite grasp what this means. I stare open-mouthed at M. Lieou who is standing in front of me. He accompanies his words with gestures indicating that all is lost. In this monastery situated nearly 13,000 feet above sea-level I had forgotten the existence of this Chinaman and his bad English. It was the only thing I had forgotten. And here I am dragged brutally back in thought into the world, my world! It appears that M. Lieou has a little receiving set in his house. . . .

The spell is broken. Two grief-stricken Europeans mourn for their conquered country in a monastic dwelling high up in the Tibetan mountains.

Outside the icy rain falls without interruption. At eight o'clock we hear the faint, nostalgic call of the shell trumpets, their despairing wail. . . .

CHAPTER 3

THE MOST BEAUTIFUL VALLEY IN TIBET

SHOTS have just rung out. We rush with Père Yang on to the balcony of the Mission House which overlooks the narrow plain of Tao Fou. About twenty Chinese soldiers, the whole garrison in fact, are scattered over the barley fields, engaged in pursuing two Tibetan horsemen whom they have already practically surrounded. The unfortunate men beat a retreat towards the river. The first one to reach it springs off his horse and, throwing away his arms, leaps headlong into the stream. That's the end of *him*; in this torrent of melted ice he'll never reach the other side. The other fellow makes a final attempt to outstrip his pursuers by galloping along the bank. But he doesn't get very far; a bullet gets him in the chest, and he falls crashing to the ground. Realizing that the game is up he throws his gun, broadsword and ammunition pouches into the water and prepares stoically for death. The first soldier to reach him draws his sabre and solemnly proceeds to cut off his head. He then heaves the body into the river and returns in triumph to the village, carrying his trophy by the hair. This bloodless head is at present hanging by one ear on a pole in the middle of the Tao Fou square. It is one of those cases of horse-thieving, they tell me. In this case the thieves were being pursued by the people of Taining. Somehow the local garrison got wind of it and managed to intercept them.

This summary code of justice is apparently the only possible one in these lawless districts where everyone is more or less a bandit. Père Yang's cook, for instance, was himself a notorious brigand only a few years ago, and whenever his master reminds him of it he puts on an air of false modesty, thus showing that he is still a bit vain about his past career. There are even more remarkable instances of this interchange of professions. The mandarin who rules the Kandze district with a rod of iron, when congratulated on his qualities of energy and courage, always remarks slyly that he learnt his trade when he was "doing the mountain". Everyone knows what he means by that.

But the mandarin of Tao Fou has always been an honest fellow, and hates the idea of shedding blood. His entourage rather deplores this tendency in him. He is a venerable old man who has done a life's service on the fringes of the Empire and of the republic. I love these old Chinese magistrates with their rule-of-thumb doctrines, honest to the core, respectable and thrifty, who, with but scanty means at their disposal, manage nevertheless to keep immense territories under control. They are the quiet servants of the law who have helped to extend their country's peaceable domination over a large part of Asia.

This good fellow is on the best of terms with our new friend Père Yang. He often comes to the Mission House and, puffing away at the cigar which he inserts in the copper bowl of his long Sseu-tchouan pipe, chats for an hour or two with the priest, asking him about the great world of which he knows nothing. Père Yang has travelled far and wide. He did his training as a missionary in Malaya, and speaks really excellent French. He is a model of deportment, this young Chinaman, and does credit to French culture.

Nevertheless, it was this charming companion who brought us evil tidings when we arrived on *June* 27. He had received letters from Tatsienlou telling him about the signing of the armistice. But these letters also mentioned something about a French Government in London which had decided to carry on the struggle. There is at its head a general whose name in Chinese lettering is Taï Ko Lo. Liotard and I rack our brains in an attempt to discover from these word-sounds the real identity of the man in whom we place our last hopes.

. . . . .

If we go on as we are going now our journey will be uneventful. Nothing of importance has occurred in our three days' progress from Gata to Tao Fou. Were it not for the rain which is a constant irritation we might be travelling through any peaceful landscape in the world. We have pursued our quiet course along mountain slopes or in wooded valleys, seeing very few houses and still fewer travellers. At night we camped in charming surroundings, the loneliness of which only came home to us after nightfall. We replenished our stock of food by shooting snow-pheasants. Yong Rine found a baby gazelle wounded in the forepaw and in a weak condition. He tried to warm it by wrapping his coat round it. But we had no milk with us, and next morning we found it stone dead.

Then, after crossing another pass, we came down into the valley of Tao Fou, which was garlanded with spring flowers. We were astonished to find ourselves riding along hedges of eglantine, and to see lilac shrubs and even wild-rose bushes near the villages, which were surrounded by fields in which the barley grew dense, high and green. This luxuriance, this richness of soil was so unexpected that it made me think of Chambhala, the Promised Land of Tibet, which lies somewhere in the north, attainable only by the few disciples of Tantrism who have acquired, by force of knowledge and virtue, the courage to face the fearful dangers which lie in wait for them, armed as they are with the ritualistic magic wand. The peasant women whom we noticed in the fields may not have equalled in splendour those creatures who, immune from the ravages of time, form the entourage of the king of this mythical country, but they were nevertheless graceful and attractive. And the men, seated on the doorsteps of their houses, appeared happy and healthy in their laziness.

Although the district between Kandze and Tao Fou is not recognized as Chambhala, it is nevertheless one of the richest districts of Tibet, and its inhabitants, the Hors, subdivided into five clans, have played an important role in its history. The Hors are said to have imparted to their fellow-countrymen their laws and methods of craftsmanship. This country must formerly have been much more thickly populated. One can distinguish in the distance, very high up on the slopes, the outlines of terraced fields, no longer cultivated, and the ruins of houses.

. . . . .

Tao Fou is a mere hamlet which has grown up beside a big lamasery whose walls, like the dwellings of the lamas, are built of clay taken from the soil on which they stand. Only the temples are of stone.

To-day there is a celebration at the monastery; the living Buddha of Kandze, chief re-incarnation of the Hors district, has come to give his blessing to the faithful of Tao Fou. The large courtyard is already filled with people, squatting Indian-fashion, facing a canopied throne. Here the holy man will hold the service, on a seat draped with rich embroideries surrounded by tankhas.

The lamas lead us along richly ornamented passages and stairs to the reception hall on the first floor, where the important guests are already assembled. The sumptuous furniture, the variegated colours on the beams and pillars, the gold frames of the glass cabinets, the lively tints of the carpets, together with the red robes of the lamas, form a truly oriental picture, the warm colours softened by the curtains through which filters the dazzling midday sun. In the golden shadows the background of mural frescoes depicting episodes in the life of Tsong Kapa adds grandeur to the polished, freshly-shaven heads of the priests as they pass to and fro in this half-mystical, half-decadent scene of magnificence.

All the notabilities of the country are there, including the Catholic missionary who has been invited like us by his "confrères". The Chinese mandarin, dressed in a robe which must have been fashionable in the Mandchou dynasty, vies in courtesy with his subjects. Our arrival complicates the situation. It is a question of who shall not drink the first cup of tea. I notice that the lamas can become very Chinese on grand occasions, although among themselves they are much more simple and unaffected and more like Europeans. Naturally, when it comes to taking our places at the window, the whole ceremonial is repeated.

Outside in the courtyard the play of colours, under a brilliant sky, attains an African exuberance. The temple, with its slanting walls, assumes an Egyptian aspect. The crowd has been waiting for hours, dazzled by the reflection from the whitewashed walls. There are people everywhere, even on the roofs; good people in Sunday dress who have come to look on at a spectacle. The scene recalls the Middle Ages, when religious ceremonies were the entertainment of the common folk.

To the sound of the trumpets the living Buddha, clothed in a golden chasuble and headed by a procession of men in plumed helmets, takes his place on the throne. Immediately the crowd bows down in rapture, their foreheads pressed against the ground. This demi-god, thirty to thirty-five years of age, healthy and robust in appearance, seems to be in no way impressed by the dignity of his priesthood. What can be the mentality of this man to whom a strange fate has allotted, since his infancy, the role of living idol? As I look at him I begin to wonder whether he is truly convinced of the sanctity of his body.

With his elbows resting on the arms of the throne, draped with a material embroidered with swastikas and dordjes, he begins by winding strips of white muslin round the heads of several priests. Then the crowd presses forward to receive his blessing. To reach him each worshipper has to pass in turn under an archway of branches supported on either side by a lama. Doubtless to incite them to true humility by forcing them to bend double, several other monks, armed with long sticks, deal blow upon blow upon the backs of the throng pressing eagerly towards this symbolic gateway. Ritualistic as these blows may be, they are nevertheless administered with vigour. The sticks, it must be admitted, are merely long, pliable switches.

Having reached the living Buddha, the man or woman, whichever it may be, bows his head low, joins his hands together and waits for this very carnal deity to be gracious enough to touch his hair. The laying-on of hands this time takes the form of a light flick on the head with a cascade of ribbons attached to a stick. His forearms comfortably supported, the Buddha raises and lowers this curious implement with a gesture which becomes increasingly nonchalant and mechanical, showing no interest whatever in the person kneeling before him. The worshipper, for his part, is deeply affected and, after the initial ceremony, moves along past a row of lamas who hold out to him every possible variety of fruit and sweetmeats in wooden bowls. Then he is presented with a khata and finally, having cupped his hands, he receives a few drops of holy water. It is a pretty and even impressive spectacle in this glorious summer sunshine. But why, when Liotard is about to take a photograph, does His Holiness see fit to burst into a guffaw of laughter?

. . . . .

Two days later we meet our living Buddha in less solemn conditions. He has been quite prosaically invited to dine, like ourselves, in the mandarin's yamen. I fancy that, accustomed as he is to a certain degree of pomp and to the luxury of the religious houses of Tibet, he must have been a little surprised, not to say pained, on entering the ramshackle shed which serves as dwelling for the representative of temporal power. In accordance with the old Chinese custom the banquet takes place at three in the afternoon. We go in the company of Père Yang, who is obviously on the best of terms with all the notabilities.

The living Buddha is accompanied by the prior and the bursar of the Tao Fou monastery. The mandarin is surrounded by a crowd of people, all related to him in some way or another. There is also a Tibetan who acts as interpreter and general factotum. The latter, Azon by name, has the distinction of being the only fat man for miles around.

Without any intentional irony in the arrangement of the guests, the Catholic priest is placed right in the middle of the lamas. The excellent bursar takes advantage of this to remark to Père Yang in the most friendly fashion in the world "that all the lamas are seated on the same side of the table". Père Yang smiles and assents without much enthusiasm. The word "Lama" in Tibetan has the generic sense of "Priest", and does not apply to any religion in particular. If one wishes to denote a Christian missionary one says "Lama piling", which means foreign priest.

The meal passes off in the usual fashion of Chinese banquets, made pleasant by the prevailing spirit of cordiality. Soon the atmosphere is enlivened by many kampas.[1] As befits the occasion, our host the mandarin relaxes to the extent of becoming slightly tipsy, at the same time maintaining his natural, unaffected dignity.

On the religious side of the table Père Yang is the only one to respond in some degree to the mandarin's conviviality. Neither the Buddha nor the lamas touch alcohol, and at the end of the meal they politely refuse the cigarettes which we offer them. But his aloof, condescending attitude does not prevent the Buddha from smiling at the sight of the flushed faces of the guests, and his smile makes me feel slightly uncomfortable.

. . . . .

Our last hours at Tao Fou were marked by a drama of jealousy and drunkenness, probably a combination of the two. A man and woman were killed and several people wounded. Incidents like these occur frequently, although the matrimonial habits of the Tibetans do not tend to excite jealousy. Alcohol, which on certain occasions they consume to excess, is often the cause of these outbursts, and the guilty ones are the first to deplore them the moment they return to a normal state of mind.

In our final partings we experienced the sadness of those who are left behind. Père Ly from Charatong came to pick up his young colleague in Tao Fou, and both of them left for Tatsienlou. It is the season when all Tibetan missionaries assemble at the bishopric for their annual retreat.

[1] Chinese custom in which two of the guests drain their glasses of wine at a single draught in token of friendship.

Père Yang's departure leaves a great blank, and we are now in a hurry to continue our journey. He has taken away with him the secrets of the outside world. With him to interpret for us we could still glean a few scraps of history from the Chinese newspapers which arrived from time to time. Henceforth we shall know nothing until the distant date of our return, that return to which we never refer, restrained by a sort of superstitious fear. The curtain has fallen before we have heard the epilogue of the tragedy which is being enacted in the Far East as well as in Europe. There is nothing for us to do but to set forth and to try and forget that in our absence the rhythm of the world is continuing as before.

Nevertheless we make a final, desperate attempt to keep in touch, and to break the months of solitude which lie ahead of us by getting news. On his return from Tatsienlou Père Yang will send a messenger after us to tell us what he has heard. This rider will have to search for us up hill and down dale and perhaps, if we are not too far away and if the tribes consent to let him pass, he will succeed in finding our tracks and reaching us. In ordinary circumstances I would not have dreamed of establishing this liaison. Once under way it is always better to concentrate entirely on one's surroundings. And having yielded to the temptation of establishing this frail link with the world, I begin to fear that the link may become a heavy chain whose weight will depress us and retard our progress—worst possible danger in a country like Tibet where one must travel fast to avoid hard knocks. And yet, try as I may, I cannot banish my sentimental attachment to far-off things, nor my sorrow at their disruption, and I rejoice inwardly at the slow progress of our new caravan with its six yaks, which advances at ox-like speed. So, as far as I am concerned, the Chinese farewell formula, "slowly, slowly", which the friendly Tao Fou mandarin shouts at us, standing on the threshold of his yamen, and watching the disorderly progress of our animals through the village, is quite superfluous. The good man accompanies his parting words with a few friendly tokens which he presses awkwardly into our hands. A sudden uprush of feeling has taken hold of him, and we feel a shudder down our backs. This old campaigner of the Tibetan frontier-land knows only too well what we may expect in the country of the Ngolo-Setas.

. . . . .

It seems that for the present fate intends to condemn us to that slow progress which I had previously so desired, and I am beginning to fear the consequences of it. Starting off with a freshly-formed caravan of yaks is no easy task. The animals at first frisk about in the grass and, when they weary of trying to throw off the pack-saddles by jumping and kicking, their thoughts turn to browsing. This wide, flat valley encourages them to stray.

The Seh river itself takes advantage of the space to meander at will over the landscape, sometimes running up against the slopes, branching off in various directions and forming little shrub-covered islets. Then it pursues its course along the base of a big landslide. Here there has been a split in the earth-crust, perhaps through a sudden, fearful upheaval, perhaps, in the course of thousands of centuries, through an imperceptible sinking of the strata, jolted by convulsions lasting only a few seconds, the more recent of which have left their mark upon the slopes. Everywhere great slabs of earth have been loosened from the mountainside, and crevasses a hundred or more feet in depth have scored the face of the mountain. The tremors which caused these scars must have been fearful.

With the rain to add to our troubles, we did not reach Charatong until *July* 31, after a four days' journey. We took up our quarters in Père Ly's empty house. We are not entirely alone in the mission. At the bottom of the garden there is the grave of a French missionary. On *March* 24, 1924, there was an earthquake and the Mission House toppled over together with all the other humble dwellings in the Christian village of Charatong. The lamas must have interpreted that earthquake as a clash of wills between the gods, because the portent was most severely felt in the very spot selected as a settlement by people who practised a foreign religion. Very few of the inhabitants escaped death. From the accounts of the survivors it appears that the wooden pillars of the church were thrown several feet into the air. The Père was killed outright in the doorway of his kitchen.

I could get no information about this Père Alric who died at the age of thirty-seven. Were it not for his name which smacks of the soil this tomb might be the tomb of any unknown Frenchman. Yet, in this remote place, it is a touching symbol. For beyond here there are no more links with France, nothing left to remind one of the countries of the Western hemisphere. A few stages westward from here, near Kandze, they found the body of an English traveller, General Pereira, who had died of exhaustion. Later on his body was taken back to Tatsienlou where he now lies buried under a rough tombstone bearing the simple epitaph: "Soldier, Explorer." Père Alric remained alone. To renew contact with civilization westward of this boundary stone we should have to make more than a year's journey across Central Asia.

Potato plants surround the tomb of our compatriot. They, too, have come from France, imported by French missionaries. And they flourish remarkably well in this soil. Forty years ago they were unknown and now they have become the staple food of the inhabitants of the district. I am not shocked to find a French Père buried in a kitchen garden. It is a custom of the Far East to place tombstones in open country, often in the middle of fields or rice plantations and occasionally in copses where people take their walks. This is not done in Tibet where the dead are never buried. Our own cemeteries, hidden away behind high walls, where social classes meet in the next world across a paraphernalia of pompous and pretentious erections, when compared with these peaceful country resting-places, strike one as exhibitions of ghoulishness and horror.

Earthquakes are not the only dangers to which Charatong is exposed. Situated on the borders of the Ngolo-Seta country, this village suffers periodic visits from these formidable plunderers from the north. Just before the Chinese revolution the Commissioner of the Tibetan frontier (the province of Si-Kang did not exist at that time) was about to start upon a big cleaning-up campaign to remove the danger of attack from these independent tribes to which travellers along the great pack-road which connects China with Tibet were constantly subjected. Chao Eul-fong, as he was called, was the very man to carry out this enterprise. It was he who, after the British expedition to Lhassa in 1904, led a troop of four thousand Chinese cavalry which reached the Tibetan capital in record time. Then, unfortunately, he was appointed Viceroy of the province of Sseu-tchouan, and left for Chengtu where he was destined to be the first prominent victim of the Revolution. When I was in Chengtu they showed me the spot where this great leader was executed. I was told that, on being bidden to kneel down, he said to his conqueror that the rules of hierarchy forbade him to bow before a mandarin of a lower rank than his own. The latter agreed and granted him the privilege of being beheaded standing up.

. . . . .

We remained in Charatong eleven days. Every member of our party except Liotard, who appears to have a constitution of iron, was laid low by a bad attack of influenza. I suffered the worst attack and for several days lay stretched out on my camp-bed with a high temperature, with thoughts and recollections turning over and over in my brain. So it was not until *August* 12 that we arrived in Luho, a little Sino-Tibetan village on a piece of jutting ground, with an enormous monastery perched above it on a ledge of rock, the whole overlooking, at about three hundred yards distance, the junction of the rivers Seh and Gni.

In order to escape the rain we took up quarters in a house where we were welcomed by a whole army of fleas. Their attentions became so unbearable that at four o'clock in the morning we left the place and pitched our tents near the monastery.

This monastery is another great foundation of the powerful sect of Gelukpas. It is a real fortified town built in square formation with six or seven different entrance gates. At all four corners of the wall are towers which contain enormous praying-mills. As was the case in Tao Fou only the temples are of stone; the lamas' dwellings are of clay. Altogether this little town, with its narrow winding alleys lined with deep gutters, has an unexpectedly Arab appearance.

There are very few people in the streets; the lamas, as in all large monasteries, are constantly occupied and never have time to relax. They employ their leisure hours—that is, when they are not engaged in public worship in the temple or in private worship in the little chapels in their houses—in domestic jobs beneficial to the community: picking wood, grooming horses, or clearing away the grass. The discipline must be pretty strict to maintain order and cleanliness in a great centre such as this which boasts a thousand monks.

At present they are busy reconstructing the kitchens. There are two enormous bronze cauldrons in the courtyard waiting to be hoisted on to stone supports. I am told that they come from Kou-Kou-Nor but, since they are at least nine feet in diameter and their weight is terrific, I cannot conceive how they have been transported all that way. Perhaps they are made of metal plates brought in parts and welded together on the spot. In any case the aspect of these cauldrons is alarming, and one wonders what potions are to be concocted in them. They are so reminiscent of the frescoes in the temples, which depict the god of Hades in various attitudes in the act of cooking men and women, stirring the fire with a pair of panther-skin bellows!

The tantric temple of Luho is another chamber of horrors, another hideous display of stuffed animals hung over the doors, of eagle-owls with powerful claws clutching the hilts of old sabres, their blades twisted into corkscrew shape by lamas in hypnotic trances. The walls are lined with repulsive masks, draped in a red, transparent muslin which only accentuates their frightfulness.

It seems that the further we penetrate into Tibet the more godly the people become. We saw in Luho a young man about twenty years of age doing the tour of the monastery on his stomach. Two steps forward, a short prayer muttered to the accompaniment of a triple movement of the hands joined first over his head, then descending to the level of his face, then to his stomach, and a final dive down with his face pressed against the earth. Performance repeated. Neither rain nor mud seemed to affect him; we filmed him, but this did not divert his attention for a moment. Any European athlete would have flinched at the physical effort expended by this young Tibetan in making a complete circle round the monastery. I have heard of fanatical pilgrims making hundred mile journeys in this style. Usually they are careful to provide themselves with leather knee-caps, for flesh and bone would never stand up to such a test.

Moreover, it is wrong to suppose that these religious practices take place only inside the sanctuaries. The lama liturgy prescribes numerous ceremonies in the open air. We stayed three days in Luho and, in spite of frequent rainstorms, the lamas remained perpetually out of doors. On the first day there was a procession. At the head came a minute sedan-chair containing a stucco god draped all over with silks and sashes; then came canopy-bearers and clarinet players followed by lamas in high, yellow felt turbans, each bearing a specially selected volume from the library, with novices and young monks bringing up the rear. The procession serpentined in and out of the curves of the ground, directed in its course by masters of ceremonies waving their wands of office. The young monks, none of them more than twelve years old, seemed smaller than the great oblong wooden volumes which they carried on their shoulders.

For two whole days the tops of the mountains were crowned with wreaths of white smoke from the bonfires kindled by the priests. Their shouts reached us from time to time, borne valleywards by the gusts of wind.

. . . . .

To-morrow we are going to leave the great Tibetan pack-trail and penetrate into the country of the Ngolo-Setas. From the monastery terrace the valley of the Gni looks like a stately V. Time after time, during our hours of waiting, our eyes have scanned this hollow, but we can only distinguish the high, regular slopes dotted with clusters of pine-trees. And already we feel sad at the thought of leaving the beautiful Seh valley with its great monasteries, and plunging into the unknown. Liotard and I take our last walk together in the dusk through the muddy alleys of Luho, rummaging about among the stalls of the Chinese merchants in the hope of finding some small object which will remind us of our civilization. All we can find is a packet of cigarettes done up in tinfoil, and we take infinite pleasure in unsealing it and lighting one of these long-forgotten luxuries. A mere cigarette is already a precious object to us; we divided the contents of this packet carefully between us as though it were some mainstay of life.

It is the hour of the evening military parade, and this too reminds us, alas, of our civilization. The Chinese mandarin of Luho must enjoy these warlike displays, for here one sees nothing but soldiers drilling. There are only thirty of them, however, and their manœuvrings are the joy of the Tibetan onlookers, who watch them with ironical detachment. As we return to our tent we hear the blast of a bugle sounding the retreat, which mingles strangely with the moaning of the shell-trumpets calling the lamas to prayer.

We reach our camp and are met by the growlings of our dog, a stray bitch whom we have adopted. We have christened her Roupie after one of Jacques Bacot's dogs. Our Roupie is handsome, but seems not to have the faintest notion of the duties of a caravan dog. Our Tibetan hirelings thought it ridiculous to take a bitch with us. Bitches, they say, are not good watch-dogs and, moreover, tend to pup, which is a great nuisance on a journey. Never mind, we like her, and hope that the solitude of the plains will revive her mistrust of strangers.

Our personnel is increased by a guide who is to accompany us for one or two stages of the journey. It was difficult to persuade him to undertake the job, for the Tibetans of the valleys detest the idea of travelling on the tablelands. However, a few days' journey to the north there is one last little Chinese garrison which protects the gold-washers. Beyond that there are only Ngolos. Once we have passed this outpost we can no longer rely on anyone's protection. And, from what I hear, travellers along the stretch of country between Luho and the district of Serba, where we shall meet our Chinese, are subject to frequent attacks.

CHAPTER 4

DIARY

The north road from Tatsienlou to Jyekundo fringes most of the way the grasslands of the independent nomads of the north, Yeh Fan (wild barbarians) as the Chinese call them. Charatong suffers from the raids of a tribe of these nomads called Seta, over whom neither the Chinese officials nor anyone else has control. Chao Eul-fong was about to take these particular nomads in hand when he left the frontier in 1911.

Sir Eric Teichman.

August 17. The thick night mist suddenly lifted this morning at the very moment when we were starting. The valley of the Gni looks peaceful and attractive in the bright sunshine. We are going to make our first step into the unknown in splendid weather.

Is it a portent? In any case it is enough to raise the spirits of every member of the party.

We make a direct descent towards the Seh. We cross it at the junction of the two rivers by a big cantilever bridge which sways alarmingly under our weight. Liotard announces the altitude—11,100 feet.

"Note it down carefully," I say to him. "We shan't find ourselves as low as this for many a day to come."

It's obvious that for weeks, perhaps even months, we shall rarely descend lower than 13,000 feet.

And then at once we enter the valley of the Gni.

"We're stepping out of the map now."

"Let's hope we shall step back into it again one day," replies Liotard, aloof and unconcerned as ever.

From now on every stride of our horses adds to our still very meagre balance-sheet of discoveries. From the earliest ages of mankind no traveller along these tracks has thought of marking out the land. We are going to submit this country to the discipline of geography.

The first stage was in no way sensational. The path follows the right bank along the varying levels of the river, across handsomely tilled fields. Just a country walk, that's all, through a charming landscape with big villages perched on rock ledges formed by stream erosions, with sunken, hedge-lined paths, and on the higher slopes of the mountain clumps of pine-trees. From time to time the valley narrows, and the base of the slopes reaches nearly to the banks of the river. Their gradient is such as to make it difficult to distinguish the more recent landslides from the scars left by the earthquake of 1924. The earth's wounds take a long time to heal.

We camp near a little village called Panda. A real Tibetan name. On all the windows is painted the sign of the Yin and Yang.

Just before halting we met a stranger wearing a curious pointed hat, and large silver rings in his ears. A man from the tablelands.

Naturally he was armed with the large double-pronged gun without which no Tibetan ever travels . . . that is, unless he has an up-to-date army rifle.

August 18. We spent a whole day here in the hope of adding something to our list of anthropological measurements before leaving the valleys. But clearly the "Pandanese" are neither curious nor trusting for none of them, not even a child, took the trouble to come and visit our camp. So, though the great Lhassa pack-trail is only a few miles off, our presence is already causing uneasiness.

Nevertheless, we were able to buttonhole a passer-by who, for two silver coins, was ready to entrust his body to us. I would willingly have given him a third coin on condition of his taking a bath, for he was astonishingly filthy, even for a Tibetan.

He reminded me of the caravaneer of whom Bacot said that he was dirty for his age because, from the fact that the Tibetans never wash, it follows that old people must be filthier than young ones.

In fairness I must admit that the uncleanliness of Tibetans seems to have been somewhat exaggerated by travellers who have wanted to set up a world record in this field. It is in any case difficult to assess the degree of sordidness of a whole population. Some types of Tibetans are quite tidy. In my opinion it is the herdsmen of the tablelands who have given Tibet its reputation for filth. These people do not even know what it is to wash, although there is certainly no lack of water in their country, which is the reservoir for the whole of Asia.

The waters of the Gni wash down a constant procession of tree-trunks. There must be an agreement between the people at the far end of the valley and the people here; the former felling the trees and heaving them into the stream, the latter lifting them out and using them for building purposes.

There is no work to do, so the time passes slowly in our little camp. Our men gossip incessantly. God, how talkative Tibetans are! Liotard and I make the most of the fine weather by taking a sun-bath. But a moment later we are hurrying into our clothes, piling one jersey on top of the other. A violent thunderstorm has come crashing down over the valley. It is suddenly very cold, hail beats down on us and our tent threatens to collapse.

August 19. Where the thunderstorm failed our bitch succeeded. Last night she wanted to gnaw her lead but instead attacked the rope which fastened the flap of our tent and nearly brought the whole thing down on top of us. Clearly we can expect nothing from that animal except to be beautiful.

It is a marvellous day. The sun is scorching. Yet whenever we pass under the shade of a tree we shiver.

Most of the women in the fields have their right breasts bared, that is when they are not stripped to the waist. And there are some girls with pretty firm breasts, wearing on their plaited hair silver cones inset with precious stones. This is the first time we have seen this style of headdress which is similar to that of mongol women. Imperceptibly, at every stage in our journey, the people change in appearance.

Those whom we met towards evening, where the houses and fields ended, were not beautiful to look at. Grimy and covered with hair, they almost made me regret what I wrote yesterday about Tibetan filth.

Moreover, we keep encountering bodies of sinister-looking travellers, armed to the teeth and wearing curious little felt hats very similar to those worn by Parisian women in 1939. Our presence on this track surprises them, but when our men tell them that we are making for Sungpan across the territory of the Ngolos they appear flabbergasted. Actually it would be much better if our fellows kept their knowledge to themselves. But how on earth can one persuade a Tibetan to hold his tongue!

And yet the swagger of our hirelings diminishes in proportion as we advance. This is no doubt beneficial to their piety which has become truly edifying. For the whole of to-day's journey they never once stopped praying out loud.

Certainly the narrow valleys along which we have been trudging for the last six hours are far from hospitable. Directly we reached a higher elevation the houses and fields became scarce and in the afternoon we saw none around us anywhere. Now we are travelling through forest-land, and forests have a bad name. It was the same in France not so very many years ago.

To get from the lower districts where the sedentary peoples dwell to the higher plateaux inhabited by semi-Nomad herdsmen we use V-shaped valleys or desolate gorges where the forest is still dense. This intermediary world, this "no-man's-land", is the haunt of outlaws and their favourite theatre of operations. Caravans always cross it at top speed. For the lowlanders it is the means of approach for those plundering hordes of the northern plateaux in their periodic raids. In earlier times they must have kept up a regular system of defence against these invasions. You can still see traces of it; enormous crumbled ramparts of earth or stone at the valley junctions. Beside our camping-place there are a few wrecked houses and the burnt-out ruins of a monastery. All that remains is a large chortain situated in a forest clearing, witness to forgotten tragedies.

Hardly have we set up camp than another caravan joins us, proceeding in the same direction as ourselves. Its occupants decide to pitch their tents beside ours. In this way there will be more of us spending the night together.

In dangerous districts travellers always tend to huddle. In former times, caravans going from China to Tibet strung themselves one behind the other, forming veritable armies which left bones and skeletons in their wake. Thus have been formed, in the course of centuries, by the continuous passage of thousands of men and animals, the deep ruts of the Imperial tracks which, with Pekin as the starting-point, radiate thousands of miles from the capital.

Our neighbours are Sseu-tchouanese who are off on a gold-mining expedition. Their general appearance at once calls to mind a "gold-rush". An entire Chinese outfit is piled on the backs of their mules: wire panniers filled with the most unlikely-looking objects, enamel basins, toothbrushes, towels hanging about everywhere. The riders are perched high on enormous poukhai and present a very unwarlike appearance. They have with them two frail, distinguished-looking women, and a lovely eight-year-old child who appears to be thoroughly enjoying the journey. The presence of these two Chinese women in this wild country is indeed startling. They have made no alteration in their dress before undertaking this rough journey, and could easily be seen about the streets of Chungking dressed as they are now.

Their escort consists of two Tibetan oulas bristling with ammunition pouches worn criss-cross over their shoulders and around their waists. These men have at least two hundred rounds of ammunition each. They keep their guns and sabres by them at night which shows that an atmosphere of mistrust reigns.

I think the time has come to hand over some ammunition to our men. For the last two days they have been demanding powder and shot like the child of Chios. Probably they're right. The guide whom we hired at Luho has already left us. With great difficulty we managed to persuade a young man to escort us a few days further, but when he realized that we intended to continue along the valley of the Gni, which made a sudden twist to the north-west, he, too, wanted to run off, without even claiming his wages. I gather that no caravan has ever dared penetrate into that narrow, thickly-forested gorge. It is apparently a haunt of the brigands of the Seta country.

I hesitated a moment and then yielded, realizing that it would be foolish to scorn the advice of our guides at the moment of our departure. I hope to find on the tablelands some track which will enable us later to reconnoitre this important river. The path we are on now leads due north, so we need only follow it and so hoist ourselves on to the plateau.

August 20. This district is obviously unsafe. Our men were uneasy this morning and wanted us to get going so as to keep pace with the Chinese caravan. But they all decamped at a very early hour, leaving us to trundle along alone. It's a marvellous day. Liotard and I are glad to be alone. Chinese caravans do too long stages to suit our own particular work.

Our yaks are reluctant as is the way with yaks, so we continue at a slow speed along the valley track which, branching up hill and down dale, leads us gradually up to the plateau. We were very surprised to find such dense woodland in a landscape usually so bare.

This virgin forest is strewn with fallen tree-trunks. It must take these giants half a century to decay. Curiously enough the bark resists better than the pulp and some of the very old ones are hollowed right out and look like huge drain-pipes.

We came up against an enormous one which had recently

collapsed right across the river, entirely barring the valley. There is no possible means of dislodging or circumventing it, so travellers have hewn out a passage through its centre. Our animals shied a bit before consenting to step through this strange doorway.

This is indeed the intermediate "no-man's-land" between the valleys and the tablelands. No one ever thinks of stopping and camping here. For a whole day we shall see no trace of a house or camp. Nevertheless we mustn't suppose that this sinister valley is deserted. We may not see anyone ourselves, but that does not mean that we are invisible to others.

So, for the first time, we take precautions. With Liotard's consent I have arranged an order of march. I and Tchrachy form the vanguard, about thirty yards ahead of the rest, then come the yaks, with Liotard, Yong Rine and Tze bringing up the rear.

Tchrachy has fixed a little home-stitched French flag on a stick into the barrel of his rifle. This emblem means nothing to the people we meet. Moreover, its strange appearance might at a distance excite the interest of thieves. But for us it is a symbol. The little flag fluttering in the breeze reminds us that we are the first Frenchmen to penetrate into these uncharted lands.

Finally, after several hours of slow ascent, we emerge suddenly into the upper world. We have escaped from the forest and the stifling gorges. Our men shout aloud with joy at seeing a horizon once more. In spite of their great experience as travellers, Tibetans are very sensitive to variations in topography. Reaching the summit of a pass is always a signal for an explosion of joy or, if they are devout, for a ritualistic observance. There is a mani on almost every pass, and every traveller adds a stone to it so that on the more frequented passes these manis sometimes attain the height of pyramids.

This country is really paradoxical. The low-lying valleys with their steep slopes, their gorges and their rock debris remind one of the Alps but, as soon as you rise above this erosion-tortured world, you come into another universe, 13,100 feet high, a world of soft, peaceful outlines like an ocean swell frozen into immobility. In summer the prairie, swept by the eternal highland breeze, is indeed that "vast green awning" which Bacot speaks of.

There should be a horde on these steppes. And by some strange coincidence here is a horde of men advancing towards us! We catch sight of them the very moment we emerge on to the plateau. They remind me at once of Huns, these picturesque barbarians, thirty in number, progressing in a disorderly fashion across the plain, driving before them several hundred yaks of varying size and colour, laden with stitched yak-skin sacks. We are face to face with our first Ngolos. They have apparently come from the Kou-kou-nor, and have already been several weeks on the road. I have at last met this tribe, dreaded above all other tribes, and I am not disappointed. These are indeed the barbarians whom Huc described a hundred years ago. They do not appear to have aged since then. And in more recent times Rockhill, d'Ollone, Teichman and others have also met them and described them. Now it is our turn. True to type, they advance upon us, on the very fringes of their territory, the territory which we intend to cross, in a line due north.

Vagabonds with long hair falling straight down to their shoulders, their ears pierced with large silver rings, dressed in sheep-skins with the wool inside, they go their way, laughing and sinister, some on horseback, some on foot, all with their right shoulders bared, bristling with ammunition pouches, broadswords and guns, the barrels disproportionately lengthened by their two-pronged forks. Some of them stop and look at us, laughing heartily at our strange appearance. Obviously they have never in their lives seen anything to match us but, unlike the people in the valleys, this does not frighten them. The covetous looks which they cast at our baggage is not tempered by any modesty.

When I see our caravan swamped in this tide of men and animals it suddenly seems to me very insignificant.

Alone once more, we set up camp for the night. Again the sense of our minuteness in this vast barren landscape comes over me. We are at an altitude of 13,950 feet and it is a bit cold. The thermometer reads fifty degrees Fahrenheit and there is a strong wind blowing. Huddled in our snug tent we listen to the silence of the night. It is an absolute silence, for the little stream which winds over the grasslands makes not the faintest sound. This silence is restful after the roaring of the torrents in the gorges and valleys.

August 21. Yesterday, when I went across to the stream to wash, I charted in the south-east a very distant snow-capped peak. If I had not made this little journey of a hundred yards this mountain would have escaped our notice. Our men loathe this deserted dell, but we have decided to override them and to spend a day here to make a tour of the horizon. If the weather is fine Liotard intends to make a latitude from the Pole star.

We are a little disappointed by our tour of the horizon. It has not given us much material. Very few of the rounded peaks which encircle the mound on which we stand are more than a hundred feet higher than our observation post. Farther off, in the direction of Kandze, we can make out one or two higher peaks, but even they have not the smallest trace of snow on them.

The snow-covered mountain mass in the south-east is hardly visible to the naked eye. We situate it in the district of Tao, quite a distance from here. It was by the merest fluke that I caught sight of it in an indentation of the hills. One tiny cloud would have been enough to hide from me its 16,400 feet magnificence of rock and ice. Thus it comes about that mountains are lost and found again. Hundreds of travellers have passed quite close to the colossal Mynia Gonka, over 24,000 feet in altitude, without any idea of its existence. Then a few people reported it, and then again for many years it was forgotten. Even to-day there are many maps which do not mention it, although it is the highest point in China.

We were surprised to discover how greatly the plateau had

suffered from erosion. The Gni and its tributaries have already eaten deeply into its surface. This country, geologically old, is topographically in process of rejuvenation. In the deep, youthful valleys, alpine forests are replacing grassland.

Everywhere else the generous outlines of the hills are further softened by their carpet of grass, which sways in the violent wind of the high tablelands.

In August, at a height of over 13,000 feet, the prairie abounds in sap. It sucks up the mass of water from the thawed peat-bogs which the frequent rainstorms continue to saturate. The vast swell of the hills is then covered with thick grass, festooned everywhere with flowers, modest or luxuriant, pale or dazzling, some of them similar to those of our own meadows, some elaborate and strange, with poisonous sap: gentians with nicely blended calyxes, deep blue forget-me-nots which attract the eye irresistibly by the simple purity of their design, mountain daisies, purple primroses, angelicas, buttercups and here and there lascivious and dangerous-looking scabious.

Man, who always tends to make himself the centre of the universe, is astounded at the luxuriance of these lands, which are so near the clouds that they seem to belong rather to the sky than to the earth. Here he must inevitably feel himself an intruder, a mere accident. He makes no roads, he does not till the soil, he builds nothing. These lands are not for him. Thanks to centuries of acclimatization, nomadic tribes manage to spend a few weeks of summer on them, but they hold on only by the slender thread of their tent-ropes and vanish again leaving no imprint but the blackened stones of their camp-fires. And if they happen to die here it is not to the soil that they entrust their remains but to the birds in the sky.

The real owners of the soil, or rather of the sub-soil, are the marmots. They are found in great masses on the plateaux, and their holes are dangerous stumbling-blocks for our horses. You see them scattered about everywhere, often in pairs, sitting upright on their bottoms, like gossiping housewives. I believe that they actually are gossiping, because they sit face to face and keep tapping each other with their forepaws, uttering little piercing cries which send our bitch Roupie into a frenzy. Roupie cannot contain herself, and makes ineffectual, bull-like charges in their direction. The strange little creatures allow her to approach to within fifteen feet and then pop down into their holes.

No Tibetan would ever think of killing a marmot. Our two hirelings must have succumbed very much to the evil influence of the town, for they enjoy taking pot-shots at them. It is true that letting off guns is the Tibetan's greatest joy, and we had to be very strict with them on this point. But to-day Liotard himself popped off one of these marmots with a rifle at two hundred yards' range; we considered that this impious act was excusable on the grounds that we had eaten no meat for two days. Tsampa and buttered tea do not agree with our carnivorous stomachs.

Marmot flesh is not too appetizing, especially when roasted on an argol fire. We have even run out of wood and are reduced to cooking our food over this dried yak-dung which gives an acrid, smoky flavour to everything we eat.

The sun had already set behind the mountain when a caravan appeared, and its leader decided to camp beside us. I am not annoyed by this intrusion, because in the afternoon I caught sight of a man watching us from a hill-top.

It is a Ngolo-Seta caravan comprising about fifty yaks. I am beginning to believe more and more that the Setas are only a tribe of the Ngolos. The five or six men who compose the caravan are dressed in hides of cardboard consistency which gives them the appearance of deformed gnomes. Only their leader wears shoes; the rest are barefooted and apparently impervious to the evening cold. All have the bold, rakish looks of seasoned trappers. I always feel slightly uncomfortable when I see these savagely armed tribesmen casting covetous eyes on our equipment. We certainly have strange neighbours to-night!

"Ya po re, ya po re," cried Tchrachy, jerking his thumb

upwards to indicate that we have nothing to fear from these travellers.

After all, one mustn't judge people by their looks, especially in Tibet, where honest men look like bandits and bandits like honest men.

This caravan, like the caravan we met yesterday, also comes from Kou-kou-nor, bringing a consignment of salt. My thoughts at once revert to the large salt-bearing caravans of the Sahara. All nomadic territories are alike.

These Ngolos have a curious way of setting up camp. They picket their animals in square formation, then form an inner ring with their equipment and settle down themselves in the extreme centre, stretched out on the grass with no other covering than their saddle-cloths.

They don't stir to-night when Liotard makes his observations. We keep switching on and off our flashlight to read our instruments. But when these tribesmen see its rays shining through the ice-cold night prudence prevails over curiosity and they leave us alone.

August 22. Almost immediately after breaking camp we meet an apparition out of the Bible, a young half-naked herdsman with a sling in his hand. David must have looked like this boy, dressed as he is in a sheep-skin loin-cloth. But David's sling was certainly not made of yak's hair. The beautifully plaited cord, about six feet in length, could be a very formidable weapon in experienced hands. Actually it is used only to direct the herds; the stone is flung out at straying or straggling animals. I am astonished at the range and precision of this catapult. Moreover, like all shepherds, the Tibetans excel in the art of throwing pebbles, even without slings. I have actually seen them killing game by these primitive methods.

Our young herdsman belongs to a large settlement which we catch sight of after crossing a ridge. This is the first tribe of herdsmen we have met since leaving Luho. Are they Ngolos? Apparently not. Our guide tells us that they are Gueschis, who owe allegiance to the rulers of Damtong a village which is the political and religious centre of the pasture-lands which fringe the Ngolo country. The sovereign-ruler of Damtong is apparently a woman. This fact is not surprising for there are numerous instances of matriarchy in this country. The great Ngolo chief is also said to be a queen.

The line of the tents of this settlement on the left and right banks of the stream appears to be almost vertical to it. This slope of the ground does not seem to worry the settlers. These collapsible dwellings are classically Tibetan, and by the length of their stay-ropes, arranged in star formation, they have often been likened to gigantic spiders. Their construction is very ingenious; unlike all other tents in the world, they are not supported by interior poles but are literally suspended in mid-air on thin props driven into the ground outside the tent; pulley-ropes are stretched over these props and fastened on to pegs hammered into the ground at a good distance from the tent. The advantage here is that the interior of the tent is empty; the disadvantage, that you are constantly tripping up against the outside ropes. The smoke escapes through an opening in the roof of the tent which can, if necessary, be covered over by a sheet. A rough wall is constructed round the bottom of each tent. Inside there is a built-up hearth something similar to those which one sees in the houses in the valleys, but apart from that no furniture. An iron pot, wooden basins, boxes of tsampa and butter, and occasionally a statuette of some god or goddess are the only things these tents contain. The people sleep in their clothes, which they hardly ever leave off, stretched out at full length on their saddle-cloths.

The tent covers and ropes are made of yak-hair. What a contrast between these black tents and those of the rich settlers and travelling lamas, made of beautiful white canvas dappled with brightly-coloured arabesques!

Each tent is guarded by a veritable monster: enormous Tibetan mastiffs who set up an infernal din, straining at their iron chains and baring their formidable fangs. I cannot help wondering what would happen if these dogs were let loose. Our gentle little Roupie keeps a respectable distance away and shows no desire to start flirting with any of them.

There are very few adult males in these settlements. The children, dressed in sheep-skin loin-cloths, watch over the herds. They don't seem to feel the cold, in spite of their naked feet and legs. Almost all of them have their right shoulders bared as in the famous picture "Les Bergers d'Arcadie". I believe that the practice, common to all pastoral peoples, of keeping the right arm bared originates from the habit of throwing stones at straying animals.

The old men, too, are very few in number, for the rough life of the table-lands is not conducive to longevity. To them is allotted the task of preparing the butter. The skin on their wasted bodies, drained of sap as their minds are drained of thought, contrasts strikingly with the distended leather of the goatskins which they use to churn the milk. Warming their old bones in the sunshine, these ancients swing the enormous leather pouches monotonously up and down, gazing absent-mindedly into space. Bold warriors once, they end their long, joyless lives in feminine pursuits.

The women are distinctly beautiful. They are dolled up like idols with silver trinkets inlaid with turquoise or coral. The loveliest part of their dress is their girdle which supports in front of their bodies a sort of two-pronged claw, a really exquisitely-fashioned trinket; in this claw is fixed the receptacle which receives the milk as it comes out of the yak's udders. Thus their principal duty is underlined by the most striking part of their attire. Drawing a parallel with our western civilization we might call it "functional architecture".

We do not remain long on the plateau. Soon we dip down again into the valleys which have slashed deeply into its surface, and which change bit by bit into wooded gorges, deserted like all gorges. We continue our monotonous descent for about five hours.

At an altitude of about 11,800 feet we emerge on to the right bank of a large, swiftly flowing river which they tell us is the Ser Khio, a tributory of the Tong. We are back once more in the valley world. The little village where we encamp is an agglomeration of stone houses surrounded by fine, well-cultivated fields. Chinese soldiers are quartered in one of these houses; it is their business to guard the gold-mines. Ser Khio, in Tibetan, means the River of Gold. In its sandy bed, as in the beds of many rivers in Eastern Tibet, are to be found grains of the precious metal for which men are prepared to fight and risk their lives.

August 23. We spend a day here putting our travel notes into shape, happy in the security of our surroundings and in the company of the Chinese. There is a woman here whose blue dress contrasts strangely with the pepper and salt of the Tibetan costume. She is a modernized Chinese, with shingled hair and a cigarette in her mouth. She is quite as alien to these parts as we are, and we admire her courage in coming to live here. But she would be astonished if we told her so because Chinese heroism is always unconscious.

August 24. For the first time in our journey we encounter really serious obstacles. The track which runs along the right bank of the Ser Khio scales precipitous wooded slopes. One of our yaks, cramped by its double load, loses its balance and rolls down the hillside. Luckily it lands up against a tree and we retrieve it without much trouble; but we find one of our cases smashed to bits with its contents scattered about everywhere. We sort them out on the spot and in doing so lose much time. To-day's journey is full of countless little annoyances and everyone is rather irritable. To cure Roupie of her abominable habit of chasing marmots we tried tying her to Tchrachy's saddle at the end of a long string, but the wretched dog got entangled among some branches of a shrub and was nearly suffocated. This little episode taught her a good lesson.

August 25. The valley has widened and we proceed quietly on our way at the normal speed of two and a half miles an hour through magnificent scenery.

We did not expect to find on our first entry into the Ngolo-Seta territory a valley so thickly populated and so admirably tilled. The harvest has already begun. The whole female population is scattered over the fields, busily scything down the heavy sheaves of corn. I am fascinated by these women's head-dress. They wear silver plaques and rings threaded through the coils of their hair from which hang long red woollen cords. Some of them, in token of respect, modestly and submissively unroll their long, black tresses, allowing them to fall down over their breasts. There they stand, facing us, charming in their embarrassment, looking us straight in the eye. Some of them have rolled down their dress as far as their waists, and their bare, bronzed nipples protrude like symbols of fruitfulness.

Perched on a rock-ledge is a monastery, in the shadow of which the female lamas are busy with the harvest; their heads are carefully shaved like those of their male colleagues. Apparently monastic regulations do not forbid the exposure of the female body, for they too bare the top part of their bodies to the August sunshine.

As in all prosperous valleys lamaseries rise up everywhere, dominating the houses of the lay population which are scattered over the fields of barley, wheat, beans, peas and garlic which constitute the richness of this lowland. Actually we are at an altitude of roughly 11,800 feet. Anywhere else this would be considered high, but it is nevertheless one of the lowest regions of Tibet.

The houses here are enormous. But their fortress-like construction gives an impression of insecurity and belies the outwardly serene appearance of this land of milk and honey. Probably attacks here are as frequent as thunderstorms. The houses are usually perched on high eminences, easy to defend, and their thick, stone walls are pierced here and there with narrow slit windows, like loopholes. They have either two or three floors, with wooden, overhanging balconies, and are topped by enormous open lofts used to store and dry the corn. So the inhabitants, even if besieged, would not die of hunger.

After the harvest season rainstorms are extremely infrequent, so the corn and the hay is in no danger of rotting.

The banks of the river are pitted with holes left by the gold-washers. It is the presence of this gold which has attracted the Chinese; we meet parties of them here and there, dressed in the national blue and with their habitual air of refinement. The mandarin in command of the district lives in the large village of Liang Po Chang which we skirted that afternoon. The presence of these Chinese seems to have softened the brutality of the natives who, a few years ago, were still undisciplined. One can only rejoice at the efforts of the Chinese to strengthen their hold over these districts which, both historically and politically, belong to them. They win over the natives by a system of gentle persuasion, at the same time making a study of their own methods of life and development.[1]

Soon after leaving the valley of the Ser we encamp in a narrow dell. Once more we are back in solitude.

August 26. The weather is too bad this morning to allow us to start off. Of course, there is nothing to prevent us travelling in the rain, but our work would suffer by it.

Our guide suddenly decided to leave us when we told him our intention of spending the day here. Why? Is it because we are far from any human habitation? Perhaps. We are beginning to realize the alarm which our presence causes to the natives who consent to accompany us.

[1] Mention should be made here of the fine work of prospecting of the Geological Survey of China, which has undertaken the task of laying down the geological map of the Si-kang, and tribute paid to the courage and devotion to science and to their country of the young scholars who have assisted in this task under frequently trying conditions.

CHAPTER 5

FIRST THREATS

The twelve tribes of the Ngolos, or "rebels", established in the bend and to the south of the Ma-Tchou, have a king who resides at Artchoun, in the valley of that river. They are a community of warrior plunderers, properly organized. Every summer they make one or more expeditions, several hundred strong, and pillage far and wide in the interests of their chiefs and of their king. No one may cross their territory except the caravans of certain monasteries and the musulmans. . . .

Fear of the Ngolos has depopulated the beautiful prairie of the upper Ma-Tchou, to the west of their territory.

Fernand Grenard.
(Géographie Universelle. Volume VIII).

August 27. It is very fine to-day, actually rather hot. We take advantage of this to make a big bound northwards.

Following the more open valleys we arrive by degrees on to a plateau, roughly 14,800 feet in altitude, where we find fresh pasture-lands and large herds. The yaks are obviously suffering from the effects of the faint Tibetan dog-star. They remain motionless for minutes at a time, their hoofs in the stream. It is the season in which they shed their winter fleece. They look very unsightly with their great moth-eaten coats flaking off and hanging down under their stomachs. Only the yearlings look nice, playing about with stylized gestures, sliding down the slopes with their tails pointed straight up in the air and butting at us with their budding horns. I am amazed at the variety of their markings. Some of them are entirely black, some black and white, some completely fallow.

Here, as everywhere, our appearance causes a sensation.

The astonishment of the herdsmen is comic, but alarming too. Clearly these men have never in their lives seen anything to match this caravan and the two men in charge of it. How will they explain us in their evening talks, beside their hearths in years to come? Shall we be men, gods or demons?

Obviously these good people are feeling the heat. The women have slipped their dresses down to their waists, and go about their business with bared breasts. They make no attempt to conceal them from us, strangers as we are.

Strangers we are indeed, the whole lot of us; Tze, our Chinaman from the Sseu-tchouen plains, and even our two Tibetans from Tatsienlou who have difficulty in understanding the herdsmen's dialect. I miss our guide: his presence might have allayed the suspicions of the natives. I don't like this fear which we seem to arouse in them. Too many exploring disasters have resulted from fear.

To-day, for the first time since leaving Luho, we had an alarm. It only lasted a few seconds, a few seconds of agonizing tension.

It was afternoon and we had just plunged into a deserted dell. Tchrachy was a little way behind, taking a pot-shot at a marmot. We were waiting for him and had allowed our yaks to browse at liberty. Suddenly we caught sight of a troop of horsemen outlined against the sky, galloping towards us from the far end of the valley. We immediately unslung our rifles, but the riders abruptly reined in their horses, turned in their tracks and galloped away. They vanished as suddenly as they had appeared.

Who were these mysterious riders and what did they want with us?

Yong keeps shaking his head and repeating "Tchapa, tchapa" at me.

The incident seems to have affected him greatly. Tchrachy takes it calmly, for he is no doubt more courageous. As for Tze, he appears entirely unconcerned. I suppose his thoughts are concentrated on his cooking, and that he is worried about the meagre fare which he will serve up for us to-night for supper. I realize his complete unawareness, and am suddenly stricken with remorse at having dragged the good man on this perilous journey.

Liotard, too, appears unmoved. But I can see by the dark light in his eyes that he feels, as I do, the vague threat which hangs over our little caravan.

Again this morning, in another of these deserted dells, we heard a shot. I went off with Tchrachy to reconnoitre. We sighted three rather surly-looking individuals with guns in their hands. At our approach they disappeared into the brushwood. What had they been firing at?

I feel uneasy. At night, in the solitude of our camp, we cannot help brooding over these incidents. Outside, our men chant their monotonous sing-song prayers until sleep eventually overtakes them. And it's not nice to hear the dog growling.

The barking of dogs expresses man's night fears. Wherever men are isolated, in mountain villages, in hamlets on the fringes of forests, in camps, there are always dogs crying out the terrors of men huddled around the fire. When, bristling up with fear, they start howling out at some unseen presence, we say: "What's the matter with the dogs to-night?" A sentence which men have repeated timorously at nightfall since the beginning of time.

August 28. If we continue northwards at our present rate we shall reach the basin of the Yellow River in under ten days. Then, perhaps, we shall locate the route taken by the German explorer, Filchner, thirty-five years ago. Since that date no one has dared cross this country, and it has remained the least known country in the world. I am eager to link the thread which we are unravelling to that of the foreigner whom I do not know and who is more than likely dead.

I feel giddy at the thought that the world's atlas is an amalgamation of thousands of routes traced at the speed of two and a half miles an hour by travellers in all ages and from all countries. France has played a fine part in this slow weaving process.

Our slow progress would be exasperating were it not for the pleasure we find each evening in tracing on the map the little line which indicates our day's earnings. For this is really how we look at it—each stage in our journey is an addition to our balance-sheet of "distance covered".

I cannot help thinking sometimes of those who have lost their lives at this game; of our compatriot, Dutreuil de Rhins, killed a few hundred miles from here, and of poor Marteau who disappeared somewhere in Tsaidam without leaving any trace.

Nevertheless we continue to thrust forward into the unknown. I must admit, however, that it does not call for very much courage. The recurring danger, its vagueness, dulls our reactions. And how peaceful this formidable country looks! Perhaps, too, because we are separated from our friends, without the smallest tie to bind us to those we love, some dark streak in our nature has come uppermost, turning us into creatures without conscience or memory, susceptible only to the baser passions such as hunger and thirst and the goal in view. If that precious rider, whom we occasionally refer to at night, succeeds in finding us in this maze, he will bring us, with our letters, unhappiness. I have often had the same experience in my travels. One line of familiar handwriting, one phrase of music, is enough to restore one's memory and power of feeling. It is perhaps for that reason that we took with us neither radio nor gramophone. Everything which reminds the traveller too strongly of the past weakens his resistance.

For the whole of to-day we remain in the upper world. Following the upper sluices of a river we reach a ridge, rounded like a dome, 15,400 feet high, which overlooks a vast amphitheatre of hills, with little silvery streams meandering this way and that. There is not a tree to be seen in this landscape, merely a few stunted shrubs lining the watercourses. The grasslands lie upon the inert swell of the barren hills like a sargasso sea.

The prairie has already lost its beauty. The grass is becoming parched, and these late-summer flowers have neither the brilliance nor the intricate designs of the spring flowers. The vast green carpet is speckled with herds. From far off one would say it was being devoured by worms. And sprinkled among this vermin are black spiders, the herdsmen's tents. Soon these herdsmen will start their slow migration towards the lower pastures, leaving these grasslands to the hard, autumn cold.

The autumn season is long delayed this year, the season so favourable for travellers, with its warm days and ice-cold nights under a finally scoured sky. The wind which sweeps remorselessly across these highlands is still chasing down the last cloud formations. Nowhere in the world have I seen them rushing along at such a speed. More and more of them appear over the horizon sailing northwards like convoys across the sea. Whenever a cloud passes over the sun the temperature becomes instantly very low.

It is indeed a strange climate. The peaks of this Olympus are perpetually crowned with fire. The rumbling of thunder is practically incessant. Sometimes the storms break over our heads with incredible suddenness, and the black squalls seem as though they wanted to strip the earth of its outer covering, whirling up great columns of dust which the Tibetans regard as evil spirits. Enormous hailstones fall round us like showers of pebbles. Then in a moment all is quiet again, like the tempest in the *Pastorale*.

We had made our usual notes and observations and were proceeding quietly along the ridge when we came suddenly face to face with a sinister group of men squatting round an argol fire, bristling with forked guns, broadswords and ammunition pouches. They sat there without moving, ominously silent. Their horses, truly magnificent mounts, were hobbled a few feet away from them.

We advanced rather gingerly towards them, but without taking hold of our rifles. They appeared quite unconcerned and allowed us to approach. Only when we had come right abreast of this curious party did one of them, a young fellow wearing the Ngolo hat, decide to get to his feet. He laid his gun on the ground in token of peace, and advanced towards Tchrachy with a swinging gait accentuated by the big broadsword which he carried diagonally across his body.

"Kanan dro? Where are you going?" he asked. His

manner was stern and he did not smile or make any gesture of friendliness.

"Northwards," replied Tchrachy evasively, and then in turn asked: "Who are you?"

The man turned his back on us and walked off without giving us the traditional "Kale kale, slowly, slowly", the farewell salute to passers-by.

We had no illusions about the business of these vagabonds. Eight men, well armed, well mounted, on this high, lonely ridge, with nothing but rolled-up blankets and saddle-bags on their horses could only be bandits.

What were they doing here? Were they waiting for us on this pass? They probably knew that only three or four faint tracks led to it, and that we should be obliged to choose one of them. Were they trying to find out which track we would take? I couldn't help connecting them in my mind with those mysterious riders whom we saw yesterday. Perhaps they'd been following us ever since we reached the plateau. Should we come face to face again with this sinister company in some carefully-chosen spot?

We were still in sight of them when they suddenly decided to mount their horses. They galloped down the slope which we had just surmounted and disappeared into a valley.

Liotard and I took counsel together. At all costs we had to reach a herdsmen's camp that evening. It would have been too risky to camp alone. As for choosing the route it was quite simple: one track was as good as another, so we chose any old valley path which led northwards.

So we began our descent.

The weather had deteriorated even more. It was cold, and there was an icy rain which soaked us to the skin. Our gloved hands began to ache. We proceeded at a snail's pace along a narrow dismal gorge, cluttered up with rocks and strewn with puddles. Our horses were nervous and our tired yaks lay down fully laden in the pools. There must have been a carcass of some kind in the neighbourhood because there were a dozen or so vultures circling over this dismal valley. We didn't exactly know where the track would lead us, and began at last to doubt whether it was a track at all. I examined the ground closely but could discover no trace of a path. Perhaps that was because the earth was so sodden.

It was a relief, therefore, to emerge at last into a valley which contained a settlement of about fifty tents. It was the hour when the herds were driven home. There were several thousand head of cattle all told—yaks, sheep and goats. To-night we could rest in peace beside this canvas village.

And so we halted, after a stretch of nearly seven hours. It was about the limit of the yak's capacity. So tired were they that, after we had unharnessed them, they remained standing motionless like statues without bothering to graze.

We remained numb with cold up till the moment when we slipped into our damp sleeping-bags.

At present it's raining, raining, raining.

August 29. In the morning we dried our things in the warmth of a pale sun. What a relief it was! Nothing is more discouraging for explorers than to get damp into their instruments. It disinclines them for work. But alas, the break in the weather was of short duration! Soon gusts of wind and heavy showers drove us back into our tents. The rumbling of thunder was so incessant that it was like the din of distant battle.

Seated on our camp-beds, our backs bent to avoid rubbing against the canvas, for the slightest touch was enough to release a waterspout, we adjusted our notes. I admired Liotard's method and perseverance. It was astonishing to see him in this deluge, manipulating compass and square with such a sure hand, marking out yesterday's route.

We established our position and were astounded at the result. We made enquiries from the herdsmen and discovered that the thin channel of water which wound through the wide valley in which we were encamped was yet another tributary of the Tong. The basin of this river is definitely much broader than we had thought. But where on earth was the Gni, which we left several days ago? We should discover it if we made a reconnaissance trip to the south-west, and our map would be so much the richer for it. Apparently we could reach it in two or three days' journey at the most. Then we should re-direct our course north-westwards, and once more continue our way across the Ngolo country.

Liotard and I agreed about this the supplementary journey. He never can resist the temptation of exploring districts which lie beyond the horizon. Never, in all our expeditions, has there been any difference of opinion between us on this point. We must find out all we can.

In the meanwhile we refreshed the inner man. For a long time now we had been on a diet of tsampa and buttered tea, made more palatable by a few dehydrated vegetables from France. Our carnivorous stomachs cried out for meat. I must confess that when we arrived here and I saw the herds of yaks my thoughts turned irresistibly towards beefsteak. But I could see we should have to content ourselves once more with a gigot of goat. A yak is too heavy a load; its quarters would weigh down our caravan.

Our men brought us a charming little kid; they insisted on showing it to us alive before throwing it on to the fire. Alas, we had no more wood and there was no tree or shrub of any kind in this dell! Try as we might to make this meal succulent, it was bound to have that stale, disagreeable flavour peculiar to all meals cooked on an argol fire. French people have over-sensitive palates. On expeditions Anglo-Saxons have the advantage over them, because with them the need for nourishment comes before the desire to eat something which tastes pleasant.

I myself have invented a dish which is the joy of my life. In exchange for a rupee or some little trumpery object, I procure from one of the tents a large pot of curdled milk. I stir it up with tsampa, cocoa powder and a little sugar. Probably I shall live to regret this magnificent concoction. All my efforts to get Liotard to taste it have been fruitless. He insists on taking his curdled milk without extras. The main thing, of course, is that he should eat it, because, being deprived of fruit and green vegetables, curdled milk gives us the vitamins we need.

Frugal as it is, our diet keeps us in good condition. We can manage our long treks on horseback without undue fatigue, and don't feel any ill-effects from the high altitude, about 14,500 feet, provided we make no strenuous effort or do any running. Perhaps it is actually due to this high altitude that I feel a great peace of mind steal over me, a kind of sublime detachment, based on forgetfulness, which comes very near to absolute bliss.

At night, huddled in my fur-lined sleeping-bag, I devote myself to reading, and live so fully in the mind of the author that I almost feel his presence at my side.

There is something exciting about this quickening of mental faculties. It is a joy peculiar to the explorer, and its psychology deserves examination; one must, of course, take account of the variety of individual experience.

August 30 *and* 31. Summer continues. That means, in this paradoxical country, that constant rainstorms prevent us from continuing our journey and that, numbed with cold, we remain seated hour after hour in our tent. At night the temperature sinks to between thirty-seven and thirty-nine degrees Fahrenheit. This would not be too unbearable if it weren't for the damp. The ground, although sloping, is as swampy as a peat-bog. We can't sit upright, with our feet dangling in icy water, so are obliged to lie full length on our camp-beds.

Our men, too, keep strictly under cover. They don't seem to be bored. The two Tibetans, having nothing better to do, crack jokes and tell each other stories. I have an idea that they're getting some fun out of being bilingual. They only address Tze in Chinese in order to poke fun at him in Tibetan. But Tze, philosophic as ever, seems not to heed their taunts. He has only one concern, and that is to get the fire going. My heart goes out to him as I see him squatting in the damp grass, trying to set light to the damp twigs which will serve to kindle the cow-dung.

At any rate we shall never want for this nauseous fuel. There are enormous sewages near the camp and we are always on the point of falling into them. Soon these mounds of yak-dung will be the only trace of the tribesmen's temporary domicile on this plateau, for the slow trek towards the lower districts is about to start and the herdsmen leave nothing behind them, not even tombstones.

By contrast with China, where the living dispute every inch of territory with the ever-growing army of the dead, Tibet is a country without tombs. One sees an occasional one here and there near the Chinese garrisons; they clash with the landscape and seem all the more sad and desolate because they are the graves of unhappy people who have died in exile. Here, in Tibet, the skies are the cemeteries. The laws of matter forbid that man should be reabsorbed into the soil without leaving any trace, so everywhere above the undulating hills float the dismal birds which feed on corpses. They say that the great gloomy creatures appear within a few seconds of a man's death, and that by some horrible sense peculiar to themselves they can discern the presence of the Great Reaper hundreds of miles away.

There is a symbolism attached to this method of disposing of human remains. And yet the Westerner cannot help feeling a deep repulsion for these sacred birds which no Tibetan would dare to slay.

Once, near a monastery, I saw a corpse dismembered. The vultures crowded round like large turkeys with a whirling of feathers, actually taking out of the men's hands large scraps of stinking flesh. They were even dealt out with portions of crushed bones kneaded with tsampa. Horrifying back-yard scuffle!

In India the Parsees have brought this process more up-to-date: they lay out their dead in the dakhmas, those famous Domes of Silence. Modern regulations have forced them to erect above these towers a system of criss-crossed netting to prevent the vultures from dropping scraps of their meal on to the neighbouring houses.

Everywhere in Asia one finds a sort of indifference to the macabre. This indifference at first astounds Europeans, but in the end their sensitiveness is dulled. I remember myself on one occasion in China going to sleep quite happily in a temple which contained three occupied coffins.

The lesser notables in the vulturine hierarchy are not so repellant. The fat, glossy crows have, indeed, a certain charm. They are arch and familiar, like all animals in Tibet. Whereas the vultures are merely concerned with the carcass itself, the crows take a keen interest in the dishing-up of the meal. In the vast Tibetan solitudes their croaking and the squeaking of the marmots are the only sounds you hear. It is a sort of ironical sneer, a little irritating to the ear. Those we got to know in 1937 in the Tsarong really seemed to be laughing at us.

Sparrow-hawks and kites are also found in great numbers. They prefer live flesh to dead meat, and their impudence knows no bounds. One day a kite actually stole the piece of meat which Tze had just put in the frying-pan, and carried it away in its claws. The Chinaman tried to revenge himself by impaling a fresh piece of meat on to a sharp dagger. But the kite was not to be had.

We have set up camp beside a huge settlement of fifty or more tents, but are not disturbed by our neighbours. There must be three or four hundred people living over there. And yet no one dares approach us, not even the young herdsmen floundering about half-naked in the swamps. This unwonted shyness is rather ominous. I prefer the importunate curiosity of the settlers in the valley who used to watch us like specimens in a zoo. I think I know why these herdsmen won't look at us even from a distance. They're scared.

Tchrachy's attempts to procure a guide have been fruitless. He managed to persuade a tall, handsome young Ngolo to come over and discuss the matter with us, but the man invented every kind of excuse for not coming. First he wanted to bring along a friend so that he need not return alone over grasslands which did not belong to the men of his tribe. Then he told us a long story about the danger of leaving the settlement where there were not enough men to resist an attack. He finished up by demanding such an enormous fee that we began to wonder whether he hadn't named this figure with the very object of provoking a refusal. This idea was suggested by Tchrachy who finally got bored with bargaining with him.

The timidity of this young warrior is unnatural. Why did not our final offer, which was a high one, succeed in dispelling his fears? I begin to wonder whether he is not more concerned with the dangers of the road than with the safety of his camp, and whether there is not some connection between the fellow's reluctance to join us and the mysterious horsemen whom we met three days ago. Our two Tibetan hirelings seem to hold the same thought, and it is with anxious eyes that they watch the Ngolo warrior's departing figure.

All the adult males of this nomadic settlement are in some degree warriors. As among other pastoral peoples, the men lead rather a lazy life. It is the women who have to do nearly all the work; they milk the animals, prepare the food, weave the yaks'-hair and mend and stitch the tents and the clothing. The children tend the herds, and the old men, scanty in number, for life on the plateaux is not conducive to longevity, swing the goat-skin bottles in which the butter is churned. The men's, the riders, duty is to protect the tribe. But as this work is not very exacting, they spend most of their day interfering in the squabbles of the camp leaders, and in the intrigues which invariably occur in tribal confederacies. Devourers of fantastic tales, the eerie, magical superstitions of Lamaism act upon them like heady wine. This is surely enough to explain the countless quarrels and violent battles which so often steep these tablelands in blood. These men, far removed from the madding crowd of humanity, with enormous territories open to them and practically no property to defend, are nevertheless unable to live in peace.

September 1. It is still raining this morning, but we have had enough of this camp; we have wasted enough time here already. There are occasional glimpses of pale blue sky in the low-lying clouds so we set off. We are now guideless for no one has agreed to come with us. Tchrachy has had the route explained to him. Apparently we shall pass many pastoral settlements, and this will make our journey less risky.

Yong Rine, however, looks terribly glum; fear is written all over his face. Tchrachy, on the other hand, younger, more intelligent and also more enterprising, is in high fettle. Whenever in the more deserted and enclosed passages I take my place at the head of the caravan, I feel the muzzle of his horse nuzzling against the hindquarters of mine.

Our route, orientated this time firmly to the west, leads us over two ridges, more than 14,800 feet high, and across rivers all of which flow south. The clear-cut outlines slope gradually down, and the classic appearance of the high Tibetan plateau stands out in its unearthly grandeur. Here the earth has not been disembowelled by erosion.

Travelling in this weather is like being in a perpetual Turkish bath. First it is very hot, then very cold. These variations in temperature are so brutal that they upset our constitutions. One moment we're sweating, and then a cloud passes over the sky or a squall of wind whips us and we shiver. The ground is sodden from frequent rainstorms, which makes walking impossible. We only dismount on the valley slopes.

This evening we discover that our bitch has disappeared. Her passion for chasing marmots has been her undoing. Roupie was untrained, and it was impossible to keep her near the caravan. The faintest squeak from one of these charming rodents was enough to set her careering wildly off in pursuit and scrambling up slopes which I had never believed a dog capable of scaling. Never once did she catch one. The other day, mad with rage, she tried to dive down into a hole hardly big enough for a rabbit. All she got for her pains were swellings on her head and mouth, which lasted forty-eight hours. I calculated that in her scamperings she covered at least sixty miles a day.

We are sorry to lose her. The European love of animals, especially dogs, astounds Asiatics. It is a form of sentimentality unknown to them. Tibetans regard dogs merely as conveniences for driving away thieves; they don't try to understand or to be understood by them. In the settlements no one pays them any attention. They are attached to posts by iron chains and remain all night exposed to nature's inclemencies; it would never occur to anyone to shelter them under a tent. I have seen them sometimes left out of doors in torrential downpours, even in snowstorms. When contrasted with this treatment, our own habit of putting little coats over our lap-dogs appears somewhat absurd.

Subjected to this upbringing the Tibetan mastiffs become regular wild beasts, capable of tearing a man to pieces. If ever one of these dogs turns on a human being the man instantly draws his sword. We found, however, that with a little exercise of time and trouble, we could always tame these monsters, and each time we found them to be treasures of intelligence and affection, more interesting perhaps than any other dog in the world. It is fascinating to discover the playful streak which underlies their rough exteriors. I knew one whose greatest joy was to chase herds of goats; no doubt he enjoyed seeing the little stilted creatures running. Actually he never did them any harm. From time to time he would make playful snaps at our horses' muzzles or try to catch hold of us by our feet. Moreover, he loved frightening yaks; the heavy grunting creatures would respond to his advances by butting at him with their horns. One day he foolishly got himself imprisoned by three or four yaks, one of whom dealt him a masterly blow, throwing him ten feet in the air. Thanks to the thick wool which lined his coat the Tibetan dog escaped unharmed.

September 2. We have apparently arrived in another territory. We spent last night in a cheerless dell, and awoke to find herdsmen's tents aligning the little river at the bottom. These men are apparently not Ngolos but Setas. It would be useless, I think, to look for any difference between the Setas, the Ngolos and the other herdsmen whom we met *en route*. It is only a difference of tribe. I believe that the ensemble of nomads in north-eastern Tibet comprises a loosely organized confederation of tribes, all of which have different names, and of which the most important and best known are the Ngolos and the Setas. Their factual independence in regard to the Chinese authorities of the Si-kang province and to the temporal power of Lhassa explains the cohesion, actually quite superficial, of this great ensemble of people, and why their territories have remained entirely unknown up to the present day.

Our first awakening in the Seta territory was painful. We are beginning to feel the effects of these last days' dampness. We are so tired that we can't find strength to strike camp; even the thought of our instruments lying outside exposed to the rain doesn't rouse us from our lethargy. But soon we find ourselves obliged to rise; a wrinkled, grim-looking man, dressed in an old, shabby red coat has come over from the camp to parley with us. In token of respect he should have let down the pigtail which he wears wound round his head like a turban; this he has not done. On the other hand he has come unarmed, thus showing his intention of negotiating peacefully. He is apparently the chieftain of a herdsmen's settlement, and his tent is about five hundred yards away from ours.

He makes an astonishing request—we are to show him our passports. We give him those which the Chinese authorities drew up for us, Our friend can't read the characters, so he stares at them with an inscrutable expression on his face, hands them back and starts an interminable conversation with Tchrachy.

The passport business was only an excuse. The man wants us to decamp. He came to us unaccompanied, lest we might mistake the approach of several members of the tribe for an act of aggression. Now he keeps coming and going between our tent and his, doubtless to consult with his subordinates, and finally returns with a curt request that we are to clear off.

We begin to feel some vague threat closing round our little band, something which we are powerless to avert. When Tchrachy tells us that the man threatened to shoot us if we didn't make a move we feel almost relieved. Perhaps we shall discover in the end whom it is we have to guard against.

At present there's nothing for it but to strike camp and resume our course westwards; the sky is clearing, but the rainclouds are still too generous with their attentions.

Just as we're moving off a dog leaps into the stream, swims across it, and rushes towards us. It is our Roupie returning in a most pitiable state. She must have been searching for us all night in the various valley settlements, for she looks exhausted and has a slight wound on her neck. Her return has a cheering effect on the spirits of us all.

After a two hours' journey, two hours of unceasing prayer for our men, we at last emerge on to a vast plateau. Miles and miles of landscape lie before us, bright in the afternoon sunshine, a vast, light green carpet across which glide swift and multiform brown stains, the shadows of the clouds. In the centre of this plateau is an alluvial plain, several miles broad, with a winding watercourse.

We see to our astonishment, in this apparently waste expanse, a gigantic edifice, a chortain bulbed like a muscovite steeple and high as a cathedral. Its presence in these solitudes is as unlikely as the presence of a ship in the desert.

The stream which waters this vast district is none other than the upper course of the river Serba which we left four days ago. Obviously we have reached the western extremity of the Tong Ho basin which we intended to explore. Beyond, much farther to the west than we had imagined, we shall meet the Gni again.

Meanwhile we halt on a ridge which has the usual stone mani bearing the magic inscription "*Om mani padme houm*"—a thousand times repeated, and reap a rich harvest of observations. Whenever, after a slow ascent like this, we reach a summit and see suddenly before our eyes a vast territory unknown to geographers, we have a sense of creation and possession which flatters our pride and uplifts us in our own eyes. These thousands of acres have, till now, existed only for those who inhabit them. Maps know nothing of them. It is we who are going to subject them to the discipline of topography.

It is distressing to think that the more we increase our knowledge the more precious our lives become, and I am sometimes frightened when I realize the dangers we still have to face before delivering our treasure into safety.

Shall we be safe to-night even, camped as we are at the foot of this chortain, facing a little chapel and three or four stone houses? About thirty lamas inhabit these houses, humble acolytes of the sanctuary which has given its name to this district: Chortaintong, the plain of the chortain.

In this lonely spot, more than 13,000 feet above sea-level, it is really an astounding construction. Bedded in a square socket, each side forty-five feet long, the bulbed tower with its gilded copper shaft rises a hundred feet above the ground. Why did the builders choose this site? What fantastic efforts they must have made to assemble the materials needed to construct this monument, symbol of Buddhism! That proud, lofty religion and the meditations of its monks are certainly well served by the majesty and cold grandeur of these surroundings. And yet the monks who gather round us are like a lot of children. At first wide-eyed and solemn, the moment they discover they have nothing to fear from us they burst into peals of inane laughter at our extraordinary appearance. Either they will try to connect us with some god or other of their pantheon and see in us some tantric manifestation, or they will regard us as freaks of nature, and we shall be the subject of their monkey gossip for many years to come.

Never in my life have I seen such simple creatures. They crowd round me and touch me, gaping stupidly at my face; then suddenly break into Homeric laughter. There is nothing awesome about my appearance and my flame-coloured hair and beard. Their guffaws of delight rather embarrass me, but they are so naïve that it would be wrong of me to show annoyance.

Liotard astonishes them less. His bronzed complexion, deep-set eyes and black hair tone better with the landscape. His curly beard alone distinguishes him from the Tibetans who are generally clean-shaven.

We learn that Père Yang's messenger came here two or three days ago. Tired of searching for us up hill and down dale, he apparently returned to Tao Fou, carrying with him the latest news from the great world. The horseman from the plains had lost heart—the perilous journey across the grasslands was too much for him. So vanishes our last hope of hearing news of France. We wanted above all to find out who was this General Taï Ko Lo who had apparently set up a French Government in London. We have no exact information and yet feel irresistibly drawn towards him.

The day ends with a grand finale of light. Nowhere but in Tibet and in the Arabian desert do earth and sky unite in such magic combinations of rare and exquisite colours. For a few seconds the chortain is like a great pink tulip; then the light fades and it becomes a large pale dot outlined against the starry sky.

Far into the night the lamas continue their ritualistic observances. We fall asleep, cradled by their baritone chanting. But our sleep is intercepted by the clashing of cymbals and the mournful cry of the large copper horns.

CHAPTER 6

CHORTAINTONG

Incidents such as these are constant occurrences in the country which fringes the grasslands of the Golok and other independent nomad tribes; our animals are never allowed to graze unwatched, and are always picketed and guarded by Domna, the mastiff, at night-time.

Sir Eric Teichman.

THIS chortain, the biggest I ever saw in Tibetan territory, was destined to mark the limit of our journey westwards. The fate which had decided to lead us to its feet seemed determined that it should be the last bright landmark on our journey. For some of our party it was to be the last human construction on which they set their eyes. It was as though this monument had released some malevolent spell, for from the moment of our first seeing it all the vague threats which surrounded us began to take human shape and, with one disaster following on another, we marched slowly to the climax of our journey.

The chain of ill-luck started the very night of our arrival. On awaking we discovered that four or five of our yaks had disappeared. Someone had cut their picketing-ropes. I remembered hearing in my fitful sleep the sound of trampling animals and even the growling of our bitch, but knowing how often before, in the witching hours, I had bestirred myself for nothing, this time I decided to sleep on. The thieves, some of whom had ambushed themselves behind a big neighbouring mani, had had a clear field in which to operate. It was a bold stroke, and could only be the work of bandits who were prepared to shoot in the event of an alarm.

Our men went off to complain to the superior of the

monastery, who was exceedingly put out by the event and offered to make a thorough investigation. We didn't set much store by this, for we knew that, however strong their moral authority, the monks could have no practical control over the vagabonds who had stolen our animals.

Nevertheless the priest's investigations had more far-reaching results than we expected. Tongues wagged and everyone talked about the affair until we eventually discovered that for several days we had been followed by a strong force of armed men who had apparently outstripped us during our last halt. Twenty-four of them all told—everyone was clear about the figure—they were evidently settled for the moment in the mountains which divide the basin of the Serba from that of the Gni. This proved once and for all that they knew our plans exactly.

So those sinister horsemen whom we had met twice in the past few days must have been scouts belonging to a much stronger force which was waiting for us in ambush, perhaps quite close to where we are now. So all the worrying events of the last week took on a sinister aspect. We had been constantly tracked down, watched and spied upon, and hostile eyes had been peering at us on those grassy slopes and blue mountains which we had thought deserted.

Now the alarm was general. Even Tchrachy appeared concerned, and Yong Rine looked green and haggard. Tze showed his concern by taking an unwonted interest in the chatterings of the people round our tents. And neither Liotard nor I deceived ourselves as to the gravity of the situation.

Our first idea had been to take the bull by the horns and to try and get past them, taking every possible precaution. There were a great many horsemen round the monastery who seemed to have no particular reason for being there. We argued that these men were probably connected with the main body, and that they had passed on the news that we were alive to the danger. Knowing us to be forewarned, our enemies might have qualms about attacking us. We had endured vague threats for so long that we felt inclined to bring matters to a head. But our men were opposed to this course. Even Tchrachy was appalled at the idea of advancing farther westwards. And yet I believe that if we had ordered him to follow us he would have done so. Previously he had always pointed out the weakness of our defence: two muskets and two revolvers against twenty-four rifles!

It was not the work of a moment to replace our stolen animals, so we had time to reflect about the best course to take. We had to send men to the surrounding settlements in search of yaks. The Grand Lama detailed two men for this errand; these men would be bound to spread the rumour of our misfortune, and also of our weakness. Ngolos and Setas would be warned for miles around and their imaginations, kindled by greed and by the orations of their soothsayers, would be running riot.

Meantime we organized a system of defence so as to avoid another theft. We divided the night into four-hour watches, without much hope of persuading our men to keep awake; we looked furtively to see that our weapons were in good order, and questioned the people wandering round the sanctuary as to the nature of the tribes of the district and their leaders.

There was a constant coming and going round the chortain. Was it or was it not due to the religious ceremonies which the lamas conducted tirelessly day in and day out?

Riders kept coming in from all sides, singly or in groups; they would make leisurely tours round the chapel and ride off as they had come, quickly lost to sight in the wide expanse of the prairie. These individuals, the pepper and salt of their yaks' hides mingling with the red of the priestly togas, did not add to our sense of comfort. With their classical pointed hats, bared shoulders, broadswords slung across their stomachs, forked guns so disproportionately lengthened as to look like lances, they were the true descendants of the barbarians of the steppes, who formerly brought terror upon the Chinese Empire. It was about thirty years ago, I believe, that they bartered their long pikes for rifles, thus depriving Western travellers of the advantage which their firearms had given them. The d'Ollone expedition, which grazed the territory of the Ngolos, probably owed its salvation to this superiority of weapons. Now Ngolos and Europeans fight on equal terms.

There has been a great deal of discussion about the advantage of carrying arms on exploring expeditions. People have said that they are a source of danger only to the travellers themselves, and that they arouse such lust of possession that instead of preventing attack they provoke it. In some cases this is true. And yet how could one travel empty-handed in a country like this, where all the natives, even the monks, are armed to the teeth? One day I saw a lama travelling quite alone, with a rifle slung over his shoulder. Tchrachy couldn't get over it: "Bad lama," he kept saying. "Lama with a gun! Bad lama!" Tchrachy is a man of the valleys.

Nevertheless, during our five fateful days in Chortaintong, I happened to see a strange traveller who carried neither gun nor sabre. He was an amiable-looking fellow, about fifty years of age, perched on the pack-saddle of a dzo, with two enormous stones bearing the inscription "*Om mani padme houm*" on either side of him. He did not notice us, but pursued his tranquil course over the prairie on his dzo, which has the ability to graze and walk at the same time. His destination was some far-distant mani, perhaps several days' journey ahead of him, where he was to lay these heavy tokens of his piety. Apart from his dzo and himself he had nothing to offer to these highwaymen.

There were also at Chortaintong fifty or more poor wights of all ages, thin and haggard, completely hairy and dressed in sheep-skin rags. When they had nothing better to do they spent their time picking lice out of their clothes, and when they found one they would go and place it on a stone, but never by any chance would they kill one. A strange country Tibet, where lice have less to fear from men than men from each other!

. . . . .

In this far corner of the desert there reigned an atmosphere of

Foire de Lendit. There were beggars, monks and peasants, every contributory element. In spite of the piety of its acolytes, the temple obviously serves as a trading centre for the district. Our arrival only tended to increase the number of transactions. Tribesmen kept bringing up for our inspection crippled and useless yaks, and we kept shunting them off again, much to the amusement of the onlookers; sometimes they would offer us the most unlikely objects, knowing that we wanted to enrich our ethnographical collection. Even the lamas entered passionately into this gold-rush, offering us hats, top-boots and even their long, red girdles. But no one ever offered us a ritual object.

Whenever we wanted to take money out of our cash-boxes we had to close our tent-flap, for greed shone in the eyes of these people whenever they caught sight of one of our glittering rupees. The value of money in districts where trade is done by barter is due more to its extreme rarity than to its practical utility. The value which isolated peoples attach to coins greatly increases the dangers of exploring, for explorers must always drag along with them a credit fortune very often greater than that of a whole district.

The yak-bargaining dragged on endlessly and we could do nothing to speed it up. The rites of transaction had to be observed in every detail. Each dealer kept repeating how painful it was to have to part with such wonderful animals, and extolling their merits to the purchasers, Tchrachy and Yong Rine, who listened patiently, making mental calculations of the animals' age. Then they squatted face to face on the grass, their wide coatsleeves unfurled to hide their hands from the prying eyes of the bystanders, and quoted prices by finger-tapping. Only after long bargaining was the price fixed; then each coin in turn had to be carefully examined, clinked and classified according to its quality, no one knowing exactly why rupees which were accepted by one should be rejected by another. This was all the more strange because the money which had been handed out to us by the great treasurer in Si-kang was silver-plated copper, a fact which all must know, because Tibetans have no scruples about splitting rupees in two to make half-ones.

To complete the strangeness of this scene round the temple we have two jugglers, one, an old man with a cruel, clever face, the other a crafty-looking youngster. Squatted on the grass, stripped to the waist, with leather gaous on their chests, they kept passing to each other a curious-looking hat, a kind of felt tiara with ear-pieces, crowned with peacock's feathers. The one who held the hat would first place it on his head, then take it off again, then hold it in outstretched arms, in varying ritualistic attitudes, chanting away at top-speed some endless recitation which the natives listen to open-mouthed; no doubt some verse-chronicle derived from the epic of the Tibetan hero Kesar who is so popular in Eastern Tibet. And then again there is a child who starts a long monologue directed at a piece of wood which is supposed to represent a gun.

Little incidents were always occurring in this strange assembly of people which threatened to develop into brawls. Whilst our men were busy bargaining, sly individuals would creep up and try to filch things out of their tent. They would take no matter what, just for the joy of outwitting us, or perhaps in order to find out how sharp we were. There was one very nasty little occurrence. Tchrachy suddenly caught sight of a man with a hang-dog expression slinking away with something hidden under his coat. He was after him in a moment, grabbed hold of him and returned in triumph with the stolen head-stall. Yong Rine in the meanwhile had seized the thief's horse and tied it up to one of our own tent-pegs. The thief took this in bad part. All three men were on the point of drawing their sabres when a group of wild-eyed strangers gathered round, preparing to intervene. We arrived in the nick of time to prevent a disaster. I smoothed the matter over by returning the horse to its owner, but realized from the look of hatred in his eyes that he would never forgive us for having caused him to lose face. He now rejoined his friends and all of them immediately leapt into the saddle and made off. A few minutes later Liotard, who was watching them through field-glasses, reported that they had halted for a moment in the plain to discuss the affair, and that he had seen them disappear in the direction of the western mountains, just where the tribesmen who were after us were said to be.

I think it was these same sinister horsemen whom we were to meet a few days later on the lonely site of a pass 16,400 feet high, and that they knew all about the disappearance of our pack animals. The danger was becoming gradually more tangible.

. . . . .

Threatened as we were, we could not make up our minds what course to take. To turn back would only precipitate the crisis; our enemies could not fail to take it as a sign of cowardice and would fall upon our caravan with greater confidence. It was possible that in spite of their success in stealing our yaks they were still frightened of us.

We were overwhelmed by contradictory opinions and advice. One day the monks came and told us that two days' journey northwards along the valley there was a powerful Seta chief who would be responsible for our safety so long as we remained on his lands. The lamas urged us to go and seek his protection. I would gladly have adopted this plan, for I was beginning to feel that only through the protection of the tribal chiefs would we escape from the bandits, and I even contemplated paying them a tribute for the right to cross their pasture-lands.

The next day (how unreliable these Central Asians are!) this exalted personage became a mere leader of a small group of tents who could do nothing to protect us.

On the other hand, they said, if we resumed our northerly course we should very soon reach, in the Ngolo territory, a village called Dekho. It was situated on the banks of a wide river, which they called the Ngolo Ma Tchou, and was the seat of a powerful monastery. There were stone houses and barley fields in this valley, and the settlers were far more advanced than the wretched herdsmen of the plateaux.

It was clear that the lamas wanted us to take this route so as to be rid of us as quickly as possible. They were obviously frightened that the Chinese garrison of Kandze, a few days' journey from Chortaintong, might take revenge on the Seta tribe if any accident happened to us on its lands. Therefore, for selfish reasons, they wanted to push us over to the Ngolos. We realized this but we had no choice. Moreover, it was becoming more and more risky to prolong our stay here.

So we finally decided to adopt this plan. We were to continue our route northwards across the Ngolo territory. In doing so we should be penetrating farther and farther into unknown parts, and increasing the distance between us and the Chinese outposts. This decision may appear astonishing to some people. But it must be remembered that we had virtually burned our boats already. It would have been useless to retrace our steps or to send messengers to Kandze for help which would probably not have been forthcoming. Even our men, let it be said, were against this course. So it was better to pursue our fortune to the end. Liotard was calmly determined on this point, and once again we found ourselves in agreement upon an absolutely vital question.

Moreover, we hoped that by this change of route we should cause our pursuers to lose time. If we made good progress we would reach Dekho before they had worked out a fresh plan of attack. Asiatics are always slow in making up their minds. We were resolved to remain in Dekho until we had bought a safe passage across the Ngolo pasture-lands. We were gambling a little on the kindness and hospitality of the lamas, and on the peaceful disposition of the farmers to arrive at an agreement.

If we were to carry out this plan successfully there was no time to be lost. We had made our decision and should have started off there and then, even if it meant abandoning some of our baggage for lack of yaks. This we did not do, and it was our first mistake. On that first error fate built up a sequence of delays which was to lead us step by step to the very disaster we were trying to avoid.

. . . . .

They passed very slowly indeed, those five days spent in the shadow of the chortain, waiting for the yaks to appear. Nature seemed to wish to add to our troubles, for the weather became appalling. Was it the close of the rainy season—we were very near the equinox—which brought all these atmospheric disturbances upon us? No two hours passed without a storm bursting furiously over the camp with a rumbling of thunder; torrential rainstorms drowned the soil, huge hailstones lashed our tent and squalls of wind threatened to carry it away, so that we were obliged to cling on to the poles to hold it in place. Added to which a damp cold froze us to the marrow and turned our hands blue.

Then, in an instant, the sky would clear and the landscape recover its olympian calm. The sun would once more shine over the yellowing hills, all of them as high as the highest mountains in Europe, yet to our eyes hardly bigger than sand-dunes in the Sahara. We would then go outside the tent and stand watching the next squall sweep the sky, like the thunder-cloud which carried Jupiter.

From time to time, during the sunny spells, the lamas would form a procession. In single file, to the sound of flutes and trumpets, they would make a slow, clockwise circle round the manis and the chortain. The higher officials, bearing dordjes and dril-bus, looked rather impressive in their clerical vestments lined with sheep-skin. Two of them wore black spectacles, probably in order to create an effect rather than to protect their eyes. They passed quite close to us, looking fixedly ahead; only the little monks who brought up the rear, urchins between ten and twelve years of age, cast sly, furtive looks at Liotard who was busy filming them with his movie-camera.

The procession would then tail into the chapel, cloudy with incense and butter fumes, echoing with music and chants, and the doors closed behind them. Then came the recreation hour when the monks spread themselves over the grass; squatting in groups they would perform in rather maidenly attitudes what most men do standing up.

Then it was our turn to become the victims of their attentions. With the naïve curiosity of all simple folk they would come and finger our belongings, making us show them the marvellous secrets of every little object; cigarette-lighters, penknives, watches. Their childish laughter introduced a gay melody into the doleful symphony of weather and landscape.

Our docile little Roupie, fastened to our tent, hardly frightened them at all. Nevertheless we tried hard to train her as a watch-dog, and she was the first to be punished for the yak incident for which we considered her in some degree responsible. This training of Roupie, as we shall see later, was to have serious consequences. It was apparently decreed that our smallest actions were to turn to our disadvantage.

On our last day we saw enormous herds of yaks departing westwards. The herdsmen were returning to their autumn pastures. So on our journey towards the Dekho valley we should find only a few belated tents on the table-lands.

Finally, on *September* 8, our caravan was complete, and we set forth in a rainstorm. We took with us an old man who had served as intermediary in the purchase of the animals. He knew the district very well, he said, and would guide us as far as Dekho. He didn't appear to be frightened at the idea of trespassing on the Ngolo lands.

This old man appeared frank and good-natured. Did he know to what rendezvous he was leading us?

That is a matter which I shall never be able to decide.

CHAPTER 7

A SEQUENCE OF DISASTERS

These nomad people, the Goloks and the Ngaba, all carry something for protection. If possible they have a gun which is always provided with the double-pronged rest hinging about half-way along the barrel; and the saying is that "A Golok never wastes a shot".
Gordon Thompson.

(*From Yunnanfou to Peking along the Tibetan and Mongolian Borders. Geographical Journal.* London. 1926.)

TO-DAY'S journey was uneventful. We were obliged to retrace our course about five miles to get back to the little dell where we had met with such a bad reception some days previously. At the top of the slope we halted, and before starting the descent, took our last look across the plains of Chortaintong. The limitless waste expanse made me dizzy, and I glued my eyes to the bulb of the chortain as to a buoy in the ocean. This was the last building we should see before reaching Dekho. We hoped to make the journey in two days, but an accident occurred which was fated to delay us a day longer.

We had already traversed and mapped down this route so, having nothing better to do, I set myself the task of preventing our bitch from chasing marmots. This sport exhausted her so much that at nightfall she would fall into a hoggish slumber. She had to be trained, therefore, otherwise she was perfectly useless to us. I had administered two severe punishments to show her the disadvantage of leaving the caravan, but in the afternoon she started off at top speed in the direction of a burrow about a hundred feet above the path.

I called to her, threatened her, but in vain. Roupie kept on running. To catch up with her I spurred my horse into a trot. Just as I was crossing a little incline in the ground the animal stumbled, fell on its knees, and threw me. As often happens at the end of a day's journey, the saddle-band was too loose and the saddle had slipped underneath the horse's belly. I got up unhurt but slightly shaken. Only an hour later did I discover that I had left my saddle-bags on the ground; they contained several precious instruments and one of my note-books.

Without a moment's delay, intending to make the most of the last streak of daylight, I set off with Yong Rine in search of them, leaving Liotard to deal with the business of setting up camp. We searched and searched, but there was no sign of them anywhere. Obviously someone had picked them up, someone who was following behind us.

After nightfall we retraced our steps. The weather was frightful. The first snowstorm of the year was upon us, and the wind drove the flakes into our faces. My frozen hands and feet were very painful, but the valley bed was so dotted about with swamps that I preferred to remain in the saddle and endure the cold.

That evening spent in that lonely valley was the gloomiest of our whole expedition. We could not help seeing in this series of accidents the hand of fate, obstinately determined on delaying our journey. A sense of despondency came over us. We should have broken this chain of prophetic events even at the cost of giving up something. But the value we attached to the tiniest item of an equipment which we had so meticulously pieced together, in which nothing was superfluous and everything necessary, forbade us to sacrifice our instruments on the altar of fate. Moreover we had been living for so long under the threat of something vague which never materialized that a sort of fatalism had got hold of us.

Therefore, at the risk of endangering our position still further, we decided to remain next day in camp in the hope of recovering our lost property. That was our second and final error.

When one reflects upon the circumstances of well-known exploring disasters (I instance the massacre of the Flatters mission, for there must be some similarity in the life of the Tibetan herdsmen and of the Touaregs) one is sometimes amazed at the apparent acquiescence of leaders of expeditions in the decrees of fate. Each time it seems that they have been driven to the crisis, at the very place where it was fated they should die, like lambs led to the slaughter. One is over-inclined, I think, to criticize these expeditions in a "wise after the event" manner and to study them like Greek tragedies, forgetting that, though the author of a play may know in advance how it will end, the actors of these real dramas who write them by their deeds, know nothing of the ultimate turn of events. Moreover, one does not take enough account of the profound change in personality which travellers experience after several months on the road. Solitary Europeans, living hundreds of miles away from their fellows, bathed in an atmosphere which, far from irritating the nerves, gives them a curious serenity of mind, see danger in softened hues, softened to a great extent by inurement but also by the capacity to forget. This is one of the most peculiar attributes of the explorer's mind, this veil, as it were, which covers his ego and dulls the edge of his dearest recollections, making an entirely new creature of him.

So, let the reader not be astonished. That night, in spite of the prescribed watches and the besetting dangers, everyone in the European tent slept as peacefully as in the tents of the Asiatics.

. . . . .

The last day which I spent in Liotard's company was a happy one.

Fate seemed to have grown tired of working against us, and Nature of punishing us. The first rays of daylight struggled painfully to emerge, snow had fallen to within a few feet of the camp, and the weather was still threatening. But as the morning progressed the whole sky cleared, and a brilliant sun appeared, a real mountain sun which quickly melted the snow. And at last it became a real Tibetan fine day, cold and dry, clear and wonderfully exhilarating.

We had been waiting so impatiently for the change of season that we were convinced that autumn was here. We enjoyed this day under our tents now dry; our blankets were spread out on the grass, while Tze busied himself piling up the last remains of his argol. We associated last week's cares so entirely with rain, wind and hail that at the first sign of good weather they disappeared; it was like waking up after a nightmare. Even the thunder seemed to have signed a truce with us.

So we awaited with confidence the return of Tchrachy and Yong Rine who, after breakfast, had ridden off to make inquiries from the few herdsmen who were still in the valley. This dell, which only a few days ago seemed so melancholy, had become gay and pleasant, and we noticed that there were still a few flowers dotted about on the prairie. Nevertheless, in spite of these happy omens, our two hirelings returned empty-handed. This was a great blow, but we didn't have much time to repine, for a shepherd suddenly arrived at the camp and told us that he knew where our saddle-bags were. His appearance, following so closely on the return of our men, was certainly something of a coincidence. But at the time we didn't worry about that. Nor were we greatly interested in the transaction which ensued, amusing though it was. The main thing was that we had recovered our instruments. The informer got a reward of sixty rupees, while the man who actually found the bags was only to receive eleven. Why eleven? It was obviously some local custom, so our men agreed without hesitation to the figure quoted.

Now that my saddle-bags were once more safely inside my tent we were able to enjoy to the full the splendour of the day.

It ended in a spectacular display of colour. The distant peaks, bathed in a golden sheen, changed from orange to purple and violet, finally melting into the blue twilight. Then from the river rose a thin veil of mist softening the contours of the landscape and making it ghostly. Only here and there, in the depths of the darkening valley, the shimmering waters of the torrent still flung out streaks of gold. Most of the tents had already disappeared, so the silence was complete.

At that moment we heard the sound of a horse's hoofs.

A man on horseback was coming swiftly down the valley. His profile, crowned by his nomad's pointed hat and the fork of his gun seemed disproportionately large in the misty atmosphere. He passed quite close to us, but did not turn his head or give any indication that he had seen us. We had time to establish the fact that he was carrying no pack. This self-effacing rider soon melted into the fog, and the sound of his horse's hoofs died quickly away.

I remained for a moment motionless, looking after him into the darkness. Where had he come from, and where could he be making for at this late hour, for the valley led only to a pass which it was our intention to cross to-morrow? This pass was obviously very high; we were still some way away from it and the altimeter already pointed to 14,400 feet. No horseman would attempt to cross it at night without some urgent reason. Perhaps, therefore, he was catching up with a party of belated herdsmen whose tents we should pass to-morrow.

I did not give the matter much further thought, and rejoined Liotard in the tent. From now on it was useless to reflect.

Since then, in my wakeful nights, the vision of that nameless rider has often appeared before my eyes with horrible clarity; the last mysterious apparition before the tragic end of our journey.

. . . . .

When, after the death of some beloved friend, one recalls in all their detail the last hours spent together, the most ordinary remarks appear significant and the simplest actions take on a prophetic or symbolic character; things which would have been meaningless had not the advent of death given them an exceptional value. Was that last evening spent with Liotard really exceptional, surrounded by objects many of which were our common property, all of which had been familiar to us for so long, in this never-changing framework of our tent, where at night everything was back in its proper place, no matter the site or altitude in which we found ourselves, for site and altitude were but external factors to the intimacy of our tiny shelter?

Now that all is past and gone, it seems to me that this last vigil was not entirely like the others. Doubtless we had come more and more under the spell of this extraordinary country, and threatened as we were by a danger vague and yet precise, hopeful from having already traversed nearly a hundred and twenty miles of unknown territory, we had reached a state of grace, an ultimate peace of mind which gave an unusual quality to our talk. Perhaps, also, we had a vague presentiment that we were approaching the crisis, and that soon our fate would be decided. Moreover, there was the sense of pride at being the first men of our race to tread the soil of the least-known regions of the globe. And the very fact of knowing that mighty events were occurring in the outside world intensified our own feeling of stupendous loneliness in the centre of this vast continent of Asia.

We talked very late that night. It was a reverie spoken aloud rather than a conversation. We allowed our minds to wander from fancy to fancy, from recollection to recollection without other aim than to express, each for his own benefit, our strong emotions, quickened by isolation and by our ascetic mode of life.

Liotard's face, modelled in light and shade against the oil-lamp placed between our camp-beds, was turned every now and then in my direction. I had never seen him smile so often, and his smile was lovely. Never had his bronzed face looked finer or more spiritual.

We talked of everything: of our murdered France, of the treason which we had begun to suspect, of the future of mankind, even of immortality. Images of loved ones, evoked haphazard in our talk, haunted the shadows of our tent.

Outside no sound disturbed the silence of the night but the chantings of our Tibetan hirelings. Gradually sleep overtook them, and now an absolute calm reigned.

Before going to bed we went outside for a moment to take the temperature. The twilight mists had dissolved, and the night was cold and clear. The sky was so glittering and the stars themselves seemed so close that we could not help feeling we were nearer to God than most other men on this earth.

But the sinister shadows of the mountains, which blanketed the firmament everywhere and through which no light shone, brought us back to reality. We ended the evening by recalling the events of the last few days. For a week or more, by a sort of tacit agreement, we had never mentioned the dangers that lay ahead. That night, however, my companion broke the pledge and expressed our hidden anxiety.

"There's nothing to fear now but an attack from bandits," he said laughing. Then he got into bed.

When I blew out the oil-lamp he was already asleep, peacefully stretched out on his back like a recumbent effigy.

. . . . .

The next morning, before mounting his horse, Liotard noted in the meteorological diary:

"90th day—*September* 10—7 o'clock—atmospheric pressure 454—temperature: dry bulb 44.6°, wet bulb 41° Fahrenheit: —ground wind N.W.—Clouds: fracto cumulus 10%—altitude 3—direction N.W.—visibility V.G."

V.G. . . . very good. That was the last remark of the last lines he ever wrote. He had initialled in advance that lovely day, *September* 10, a day which from dawn to dusk was to be superb.

The route was easy. We advanced up the valley, following the winding stream which became gradually shallower, over a spongy, springy soil, still grass-grown and dotted with flowers opening out in the sunshine. The marmots were warming themselves beside their burrows, and the noise of our horses' hoofs sent these little rodent mammals of the grass-lands scampering in all directions. The prairie was celebrating.

Tchrachy had recovered his spirits, and was singing at the top of his voice; the same old song he had always sung from the first day we started; a great piercing call ending with a long drawn-out note which echoed across the valley.

The valley was deserted. But far ahead of us there was a melancholy roundabout of vultures describing their concentric circles over something which we could not see.

Our course was slow. The yaks strayed over the grass and had to be whipped up from behind. The ones we had just bought were unused to pack-saddles, and kept trying to shake them off by rolling on the ground or making sudden mad rushes. Our guide, trotting along on foot, urged us forward, in order, he said, to reach Dekho that evening, for the journey was a long one.

After an hour's march we came upon two tents, right underneath the parade-ground of the sinister birds. One was yellow and entirely occupied by lamas, the other black, a simple herdsman's dwelling. In the black tent was a dead man. The lamas had come to camp near-by in order to perform the funeral rites of distributing the remains. Beyond there was nothing at all, and the path was hardly visible. We had crossed the fringes of the highest summer camping-grounds, and were about to enter a world where nothing lives.

Nevertheless, we made one more contact. A gold speck on the landscape heralded the approach of a lama who was coming down the valley accompanied by three servants and saddled mules. He was obviously a high official of the church for he wore the gilded cap whose only function, I believe, is to impress the highwaymen with its wearer's importance. Our men dismounted and did obeissance to him. But instead of returning their civility with the graciousness common to priests of every religion, he questioned them rather abruptly about the route we intended to take. The good man seemed nervous and preoccupied, but this did not impress me very much at the time.

"Don't go up to the end of the valley," he said. "The way is very bad there. Take the first ravine you come to on the right: that also leads to Dekho."

He was already several yards behind us when he leaned back on his richly-caparisoned steed and called out sternly:

"Take the right road—the right!"

About an hour later we reached the place which he had indicated, and instead of turning to the right we turned to the left. Why? Just from instinct? Because the ravine was narrow and the valley wide? I can't explain. There were five of us, a Chinaman, two Tibetans, and two Frenchmen, and none of us uttered a word in protest. It was the guide who decided. He alone, probably, knew the real reason. And yet we didn't have any particular confidence in him nor any mistrust. We didn't even bother to question his decision. We were progressing and that was enough.

As for the lama, he had done his duty according to the Tibetan creed. He knew exactly what was waiting for us on the pass. But if an unrelenting law of causality was to lead us there, it was not for him to interfere between cause and effect. Nevertheless he had indulged himself to the extent of offering us fate's warning: he had acquired virtue by doing a good deed, and had departed content.

Thus we continued our slow ascent towards the rocky crest. So slowly did we advance that the guide told us that we should not reach Dekho that night and that we should have to pitch our tents in the mountains. At last, shortly after midday, we caught sight of the pass at a turning in the valley.

CHAPTER 8

THE LIOTARD PASS

In the country they are known by the generic name of Kolo (Ngolo). Their retreat is said to be hidden among mountain gorges which are inaccessible without a guide. The Kolos only come out to scour the desert and to indulge in pillage and devastation. It is said that these brigands have the revolting habit of eating the hearts of their prisoners with the object of maintaining and fortifying their courage. Moreover there are no monstrous practices which the Mongols of Kou-Kou-Nor do not attribute to them.

R. P. Huc.

Recollections of a Journey in Tibet. 1844.

LIKE all other passes, it was a niche of sky seen through a barrier of bare, jagged rocks; but, as though it were in some way accursed, there was neither mani nor chortain on it, not even a prayer-banner. Obviously no traveller had ever dared cross it. Nevertheless a half-obliterated track led straight upwards to the ridge; the track of some migratory herd, or perhaps of bandits too hard pressed to observe the ritual.

We advanced slowly, our eyes fixed on the bright opening. In spite of the threat which hung over us, we did not feel much uneasiness. Crossing a pass is, of course, always a risky business; one can fall into an eternal sleep, numbed by cold or fatigue; one can be killed by man-made bullets; moreover, lonely heights are always haunted by demons. No one can say which is most to be feared—nature, man, or demons.

But these passes! We had already crossed so many of them! And it was such a lovely autumn day, so clear and dry! We were carefree, like nomads who have left their sorrows several days' march behind them and, in the full knowledge that sooner or later they are bound to be attacked, nevertheless continue their endless journey.

However, when we reached the base of the rock barrier, our Tibetans became solemn. Their full-throated singing changed to a low monotonous litany. They only stopped their incantations to shout encouragement to the animals, whose progress was impeded by the steepness of the slope and by the thinning air.

These murmurings were the inevitable accompaniment to all dangerous crossings. They were always the signal for me to spur my horse forward to spy out the land, and each time Tchrachy would follow close behind me.

We advanced sedately, I looking to the left, Tchrachy to the right. Our horses progressed in jerks, stopping to take breath after each effort.

This time, too, we saw nothing but enormous boulders split by the frost, strewn about over a dead ground. But when, with a final straining of their hocks, our horses hoisted themselves on to the windswept ridge, we suddenly beheld a vast panorama.

The Ngolo country lay before us in all its splendour. Never, perhaps, in all my travels, had I felt more strongly the pride of discovery, the acute joy of spiritual possession.

"16,100 feet," announced Liotard laconically. He began stuffing his pipe with tobacco, as was his habit before making his observations, and his face revealed nothing of his inner feelings.

And yet it was indeed moving, this vast landscape which we were the first Europeans to set eyes upon. Between thirty and forty miles to the north-west, scarcely visible to the naked eye, was a range of mountains covered with snow and glaciers. We examined it through our binoculars and saw that it was a tangled formation of boldly-jutting peaks, crowned by an enormous fortress-like massif. The snow and ice which covered it proved that it was a range of great altitude, because all around us and on the steep rocks which rose about two or three hundred feet above the pass, there was not the smallest trace of snow. Without overestimating, we could judge those mountains to be 19,500 feet high. They could be none other, therefore, than the Bayen Kharas. The terms of our mission immediately leapt to my mind: "The object of the expedition is to establish the site of the sources of the Tong and to locate the eastern limit of the Bayen Khara mountains. . . ."

Between them and us there was no trace of the plateau upon which we had spent the last few days. We had reached the edge of it. On the other side of the pass there were only the remains spared by the powerful erosion of some unknown river which might well prove to be the Tong; a jumble of grass-grown ridges interspersed by deeply-hollowed valleys, from which emerged here and there the fringes of the forest.

In the whole of that immense space there was nothing to indicate the presence of man. Scan the landscape as I might, I could see no house, no monastery, not even a prayer-banner. The Ngolo country seemed to be deserted. I confess that I would have preferred to see a plateau dotted with tents, and that the prospect of plunging down into the narrow trench which lay open before us, and where our view would be barred by the overhanging crests of the slopes, made me lose heart.

While we stood there feasting our eyes on a landscape which we should never see again, the caravan had assembled on the narrow bridge of the pass.

"What shall we do with the animals?" said Liotard. "It's already late, and if we don't want to camp too high we've got no time to lose. If we let the yaks go on ahead we could easily catch up with them on our horses and thus gain time."

"Oh, what does it matter," I answered unthinkingly. "A few yards higher or lower won't make any difference. A short rest won't hurt anyone."

What would have happened if our caravan had continued its journey? The fate of all of us might have been different. For the second time on that memorable morning we had been offered an opportunity of escape. And for the second time we let it slip.

So we began our usual tasks: photographing, sighting, jotting down things in our note-books, examining the landscape through binoculars. . . . The binoculars revealed nothing, absolutely nothing to cause alarm.

I recollect that Liotard had to lie flat on his stomach to make his observations, as the wind was so violent.

Meanwhile, a few feet away, the caravan was waiting. The Tibetans had followed Liotard's example and lit their bamboo pipes; they were gossiping, with their hands on their horses' bridles. The horses never neighed once. There was nowhere for the yaks to graze on this dead earth, so they stood there motionless like statues. Our dog Roupie, crouched like a sphinx, was relaxing after her day's scamperings after the marmots. During the whole of our ten minutes on the pass she didn't bark once, she didn't once growl. She remained quiet, absolutely quiet.

At length we resumed our march. Just as I was plunging down the left slope of the valley, Liotard, laconic as ever, remarked:

"16,250 feet."

He had just tapped the glass of his altimeter with his finger to set it going.

That was the last time I was to hear his voice.

The sky was marvellously blue and the rocks of the pass shone fawn-coloured in the sunlight. Down below us, at the foot of the valley, a little stream meandered across the yellowing grass-lands. My watch said 12.30.

. . . . .

Everything that followed is engraved on my memory in a rapid succession of pictures, a sort of uninterrupted film-sequence of violent scenes.

We had just started our descent. I was walking at a good speed down the faint track which ran midway along the slope of the hill. My horse's bridle weighed heavily on my arm, and I felt its muzzle pushing gently into the small of my back. Tchrachy was on horseback, a little to my left, his rifle slung over his shoulder. I could hear him praying.

After a few steps I turned round to make sure that the caravan was following in good order. The ten yaks were barging about on the narrow ledge, knocking their loads together, panting and snorting. Behind them Yong Rine was cracking his whip, urging them along with shouts. Tze, placid as ever, was leading his horse by the bridle and paying no attention at all to his surroundings.

Liotard, on foot, brought up the rear, leading his white horse by the bridle. Handsome, serious and black-bearded like a stained-glass window saint, he wore that familiar expression on his face, cold and contemplative, sunk in his own thoughts. Like me, he was probably wondering about the country that lay ahead of us, picturing those far-distant mountains, contemplating the future.

How extraordinary that at that moment none of us, not one, should have had the faintest shadow of doubt or premonition!

And yet two of us were within a few minutes of death.

I had just seen Liotard for the last time.

. . . . .

A shot rang out on my left, a little behind me! It was so muffled that I would have thought it a long way off, had not a sound like the buzzing of a bee told me that a bullet had just grazed my ear.

For a split second I thought that the shot had been fired by one of our party. I turned round in astonishment. There, behind a rock, was a man with a gun levelled at me. The barrel was still smoking, and he had not yet had time to re-load. He was so close that I could see his dark eyes fixed on me.

So that was that! We'd been attacked. The first shot was the signal for a volley. Later on I was to discover that the bullet had pierced my jacket in two places.

For a moment I considered taking cover behind my horse. But the poor beast had been hit and started bucking. I let go of his bridle. He was probably fatally wounded.

In a flash I saw several more Tibetans ambushed behind rocks, forming a kind of redoubt. They were firing at us almost point-blank, and in a circle of twenty feet radius had complete command of the narrow ledge where our caravan was assembled. They'd chosen their position well. They had us entirely covered and, to make their position still more secure, had waited for me to pass before starting the fight. The result was that Tchrachy and myself were cut off from the others both by their guns and by the frightened yaks who blocked the pass.

If I stayed three seconds longer on the track I should be riddled with bullets. There was nothing for it; I had to jump down into the valley at the risk of breaking my legs. So I hurled myself down on to some scraggy bushes which broke my fall, and at the same moment saw Tchrachy leap off his horse and seize hold of the sling of his rifle. The agonized expression on his face proved once and for all that there had been no treachery on the part of our men.

My hands were torn and bleeding. I rolled down a few feet and landed up against a boulder. Here I could take cover and return the enemy's fire. Lying on my stomach on the sloping ground I took my revolver out of its holster. I should have to shoot high and the peak of my pith helmet got in my way. For a moment I considered throwing it away, but then a confused thought occurred to me that it was a mistake to dispense with anything in a crisis like this, so I calmly fastened the strap under my chin and threw the cap on to the back of my neck.

I let off my first shots haphazardly. I was shooting from below, so I couldn't even see where my shots were going. I should have been extraordinarily lucky to have hit even one of those men, and I continued firing more from principle than anything else, to show them that we meant to defend ourselves.

Now that I could see the situation as a whole, it was obvious that we were in a tight spot. Our attackers had the enormous advantage of position. The intensity of their fire proved that they were also superior in numbers. There were at least twenty of them, perhaps more, and I guessed them to be bristling with ammunition pouches like all bandits. We, for our part, were four armed men; we had two rifles and two revolvers between us and very little ammunition. It was useless to think of our ammunition cases; they were attached to our yaks' backs and therefore out of reach.

If only we had been all together! But I didn't even know where my friends were. They must have taken cover somewhere so as to return the fire. And I could actually hear them shooting back; the reports from the European rifles were much sharper and the clean whistle of their bullets unmistakable.

They were shooting! They were alive! That was all that mattered for the moment. We were in a jam and our only hope was to unite and escape from this massacre.

Our attackers kept up their running fire. I could see their faces above me and the barrels of their guns resting on their hunting-forks. In the rarefied air of this site, higher than Mont Blanc, the noise of the reports was extraordinarily muffled. The bullets ricocheting round me sounded much louder than the actual discharge. Our enemies' powder couldn't be too good because the whizz of their bullets was weak and sizzling. I had the queer impression that it took them some time to reach their objective.

Even more remarkable than the sizzling bullets was the silence of the shooters. Since Liotard last spoke I hadn't heard a human voice. Our enemies hadn't made a sound, not even at the very first moment of the attack. No wild yelling to express their hatred and frenzy! They just went on mechanically shooting, loading and unloading their guns with the diligence of conscientious workmen. There was something pitiless about this silence. What could it mean? Usually Tibetans are more noisy in an affair of this kind. When they pounce upon a caravan they rely as much on their shouts as on their weapons to terrify their prey. Why were they so quiet to-day? Why were they not revelling in their war-cries? I was beginning to feel vaguely that this time there was more in it than mere plunder.

Yes. There was certainly something more in it. From the very beginning our enemies had had the advantage and yet the firing didn't stop. They were pretty well assured of their booty, and yet they didn't seem to want to parley.

So it was our lives they were after! They'd been following us all the time, not to rob us but to kill us. In that case we were as good as lost. In the waste of these desert mountains we couldn't expect any help. Even if by good chance travellers were to happen on this battle, they would take good care not to interfere; they would at once turn about and take to flight. And left to ourselves, I did not very well see how we could escape.

So I was very probably going to die. It was funny how the idea suddenly seemed so ordinary. Perhaps I was realizing for the first time that life ended in death. Earlier or later, it made no difference . . . and yet I knew very well that I loved life, that I was tied to it by a thousand chains which bound me to other men, men who at this moment know nothing of my plight, who would never see me again and who would suffer. Several months hence they would hear the news; they would not understand how simple it was to fall back on this faded grass and never rise again; they would make a tragedy out of an ordinary, everyday misfortune, the summary execution of a few travellers on a mountain path. Just a banal adventure, very banal really in this gloomy dale, which seemed to have been created purposely as a setting for it.

I had to do something so I went on shooting; for the moment no other plan occurred to me. The buzz of those pestilential insects passed so close to my ears that I hardly dared move my head. In any case, where could I go? One place was as good as another so far as I was concerned.

How long had I been lying behind this rock? I seemed to have lost all sense of time. I knew that I had already emptied one clip of my revolver and inserted another. I had also unsealed my last two packets of cartridges and shoved them in my pocket. All that must have taken quite a time, too long probably, because now there seemed nothing more to do.

I was now counting my shots, resolved to keep the last bullet for myself through an absurd fear of being polished off with cold steel.

. . . . .

I was suddenly startled by a noise of crackling undergrowth on my left. It was Tchrachy, bare-headed, with a tense look on his face, clambering down the slope at a mad rate, dragging behind him an unsaddled horse. The horse kept sitting down, getting up again and sliding stiffly along the rocks with grotesque movements. I gazed in agony at this ludicrous spectacle, sensing with every nerve the invisible network of bullets which surrounded it, expecting every moment to see man or beast collapse suddenly in a heap. And yet there they were at the bottom of the valley, near the spring, and Tchrachy had taken cover behind a rock. Only then did I notice that he had lost his gun.

Was that the end already? But why was Tchrachy the first to take flight, and why, in that case, did he remain there, when it was only a question of putting spurs to his horse and riding off?

Oh yes, it was the end all right! Much farther off I saw a man slithering in and out among the rocks with his gun in his hand. Yong Rine, in his turn, had left the rear-guard. The realization that the only two men capable of parleying with these savages were deserting us removed my last hope. A sense of panic seized me at the thought of being left alone in this desolate country, stronger than the physical fear of death, and I instinctively clutched my revolver, my last support, my most precious possession.

No! Our Tibetans were not fleeing after all. Yong had joined forces with his nephew, knelt down behind the rock and fired two shots in quick succession.

I, alone at the head of the caravan, clinging to the slope, thought of Liotard also alone, only three hundred yards away from me. Old Tze couldn't be much help to him. Poor man! How was it he hadn't fled? And our guide? He must have run off ages ago. That man's treachery didn't occupy my thoughts for long. It didn't matter now in any case.

Liotard was fighting back; I was sure of it. Otherwise the firing wouldn't have been so intense. I fancied that the reports from his revolver were more distant. Perhaps he was trying to climb back on to the pass and take cover on the other side. But the ridge must have been entirely exposed. He should have followed Yong. Perhaps it was already too late and he hadn't had the choice. Poor Louis; it was silly for us to die like that, just a few hundred yards away from one another.

It was a ridiculous situation altogether. I couldn't join up with Liotard without running the gauntlet of the enemy's fire, and, where I was now, I had nothing to expect but a bullet in the head. I was completely cut off, and if the enemy took up different positions, they could get me from all sides. What was I doing here? I had to move quick, and not having the choice any more than Liotard, I would try and join up with my two men. It was a risk worth running. I should be able to run better if I took off my leather jacket. So I unfastened it and slung it over my arm. Then I started running. . . .

The devils! They were shooting at me like a rabbit. The horrible whizzing pursued me, and I felt as though I were tacking about between invisible metal bars which at any moment I might run up against. I ended my course with a masterly tumble which landed me at the bottom of the slope. Perhaps that tumble had saved me. At all events, when I got up, my heart was thumping and my lungs bursting. How I cursed the altitude! If I made any further effort I should collapse and just remain lying there. The Tibetans might be bad shots at a moving target, but they made very good scores on a stationary one. I had still thirty yards to go before reaching the cover. Nevertheless I took it slowly, crouching, my head bent double as in a sandstorm, with bullets raining round me. . . .

My men watched me coming with amazement. They didn't understand and no doubt they thought I had gone mad. The Tibetans have no idea of the sickness which we Europeans suffer on these heights.

There was a nasty surprise waiting for me behind the rock. Tchrachy was lying full length on the ground. He looked up at me, his face pale and distorted. He was wounded!

At first I could see nothing but a slight scratch on his chin. That bullet might easily have shot off his jaw. Then I noticed his torn clothes. He kept trying to tell me something which I couldn't understand. At last he touched his back with his hand, lifting his forefinger in the air and bending it double to indicate something. I got his meaning. He was trying to tell me that his spinal column was broken. That complicated our position appallingly. If by a miracle we managed to escape, what should we do with him? But, on thinking it over, I realized that people don't run with splintered backbones, and I kept shouting at him, that he was wrong, wrong. I really think that at that moment I began laughing like a madman.

Tchrachy was really very brave. In spite of his wound he remained astonishingly calm. Yong Rine, however, was green in the face and trembling all over. Nevertheless, lying flat on his stomach behind the rock, he kept his rifle to his shoulder. The brigands were still firing at us, and I could see in his eyes that he was expecting them at any moment to come rushing down on us.

And Liotard? I got up behind the rock and tried to catch a glimpse of him. Where could he be? I was actually beginning to feel annoyed with him for not having followed Yong. Stare as I might, I saw nothing but drifting smoke from the muzzles of the rifles still pointed in our direction.

"Yong, where is Lieou Sien-Cheng? Tell me, where is he?"

In my excitement I took him by the shoulders and shook

him. Only a few minutes ago he was with Liotard and Tze. He must know.

But it was just a waste of breath. Yong was mad with fear and his teeth were chattering. My first thought was that he had seen them killed. But then he pointed towards the pass as though to say that they had retreated over the other side. Was he lying? Was he out of his mind?

Then suddenly an ingrained racial hatred made him open his mouth, and he poured out a string of oaths against our Chinaman. Apparently the poor chap had put up no defence at all. I gathered it was all up with Tze. Yong made a horrible gesture to indicate that the man would never again see the peaceful, fertile plains of his native land. I learnt afterwards, when Yong Rine had recovered his calm, that the wretched Tze had made no attempt whatever to defend himself, either by taking cover or by using the rifle which we had given him. He was shot standing up, and fell back dead against the rocks, probably without even knowing why he was being attacked.

I couldn't believe that Liotard had also been killed. I was certain now that I could hear above the shooting some sharper reports from farther back. That meant that he'd retreated and that the bandits were following him. His fate and mine were about to be decided. His chances were as good as mine and my men's. But, of course, there were three of us together. He was alone. . . .

The brigands had probably separated into two bands, one following Liotard, the other to deal with us. To speed up the slaughter, the second lot was beginning to deploy into sniper formation, so as to get us from behind. They wisely kept to the heights. None of them dared approach us. They'd have had to come into the open, and they didn't like the thought of facing our rifle, the only decent weapon we possessed. We shouldn't even have the pleasure of bagging one of them. This timidity on their part gave us a ray of hope. But if we stayed much longer where we were we should be completely surrounded. And, once surrounded, we would hold out as long as our ammunition lasted and then finis! So we had to get down to the bottom of the valley before it was too late.

Thank God Tchrachy had had the presence of mind to save a horse. We helped him hoist himself on its back. Oh, the prestige of the overlord in this feudal land! The wounded man hesitated in my presence to put spurs to his horse.

Tchrachy disappeared at a trot, lying with his head on the horse's neck. And then we got ready to leave our precarious shelter. To scatter the fire Yong started off ahead of me. I envied the cat-like agility with which he leapt from rock to rock, only remaining a few seconds at a time exposed to the fire. My own lowlander's heart and lungs wouldn't allow that.

Moreover, I felt an overwhelming fatigue steal over me. I was beginning to realize the full horror of the disaster. So much effort, so much toil, and all to no purpose! I felt suddenly that life and death were no longer very important. So I started off slowly, dangling my revolver in my right hand, with bullets still whizzing round me.

I suddenly realized that the men in ambush above, seeing me walk so casually and ignorant of the effects of altitude, must be thinking me invulnerable.

I walked on automatically, looking at the sparse shrubs which were the soil's only adornment. For the first time I realized how ugly they were. I suddenly felt cold, and instinctively raised my head to look at the sun. A small white cloud passed over it, and then continued its quiet course. The sun must have been already low in the sky. I wondered what time it was. I could easily have found out by looking at my wrist-watch. But there didn't seem any point. . . .

Gradually, almost imperceptibly, the shots became less frequent. By the time we reached Tchrachy, who was waiting for us out of range, the firing had ceased altogether. Our attackers hadn't attempted to follow us. They hadn't even dared face our two poor little weapons.

Everything was quiet now. That meant there was no more fighting. That silence terrified me.

We stopped and counted our ammunition. I was furious to discover that, in my fall, I had lost all the ammunition which I had stuffed into my pocket. I only had four more shots. Yong Rine was even worse off. He only had a clip with three cartridges. Seven shots in all! So to all intents and purposes we were unarmed. We were helpless, and if our enemies decided to come after us, we shouldn't escape.

Then, just as we were examining our guns, we heard a last, single report in the distance. My two men looked sharply at me, not daring to speak; then we all stood gazing at the pass, which had resumed the mournful aspect of all lonely, desolate spots.

A few seconds passed. Then, deadened by the distance and by the thinness of the air, came a sound of wild shouting. And then again nothing but the great silence of high mountains.

We remained motionless, our eyes fixed on the dismal rock barrier outlined against the bright sky. We could just distinguish in the distance a black spot which we realized was our troop of yaks, and a white horse grazing in the brushwood. It was Liotard's horse.

CHAPTER 9

THE DISCOVERY OF THE TONG

This vast extent of country is inhabited by the Golok and other wild nomad tribes, and, being a closed land to Chinese and foreigners, remains one of the least-known parts of Asia. It is enclosed in a rough triangle formed by the following three caravan routes: Tachienlou Kandze, Kandze Jyekundo, Jyekundo Sining, Sining T'aochow, Sungpan Tachienlou; all these roads are subject to raids from the nomads living inside the triangle they enclose. The Golok appear to recognize the temporal authority of the Lhassa Tibetans as little as they do that of the Chinese. The possession of rifles, which they acquire in the course of trade from the Mahometan merchants of Kansu has made them far more formidable than they used to be.

Sir Eric Teichman.

THE poor scattered remains of the caravan formed a miserable little group at the bottom of the valley. At present we were out of danger, but our enemies had so clearly revealed their intention to massacre us all that we were justified in thinking they would try and pursue us. Like all Tibetan bandits they would be well mounted; we hadn't seen their horses, but they were probably tethered some way away from the pass so as not to reveal their presence by neighing. It was no use even contemplating returning to the scene of the tragedy. In that desolate spot the tribesmen had no cause to hurry. They could go quietly and methodically about their business of dividing up the spoils. They might even wait till nightfall before dispersing. We should be completely in the open, and our every movement visible to their sentries.

Our first idea was also to wait till nightfall and then make the ascent to the pass. In spite of Yong Rine's insistence that Lieou Sien-cheng had taken cover on the other side of the mountain, I was mortally afraid. That last report still rang in my ears like a *coup de grâce.*

But if we took this course it would mean waiting for hours. I had at last looked at my watch. It was two in the afternoon, and it would be dark at seven. Moreover, the wounded man was in great pain and needed attention. We had nothing left now, no drugs, no provisions, not even a container for boiling water. Moreover, we were far above the zone of conifers which could have provided us with torches. And without torches how would we find Tze's body in that maze of rocks? And even if we found Liotard alive but wounded, what could Yong Rine and I do with two wounded men and only one horse to carry them? In the icy cold night, without food, without alcohol or any hot drink, we could do nothing for them. We should have to leave them and go in search of help.

The last tent we saw was the one with the dead man whom they were preparing to sacrifice to the vultures. It was almost three hours distance from the pass. And I doubted very much whether we should still find it. Our guide, in his flight, had probably warned the herdsmen, who had then taken to their heels. And beyond that what? Only Chortaintong, which was a good day's journey from here.

Dekho, on the other hand, which we were making for at the moment of the attack, was apparently five or six hours away at the most. We had a good chance of reaching it that evening, and of organizing a caravan that very night.

Virtually disarmed, exhausted by a long day's march and by the unequal struggle which we had waged for nearly an hour, we had the choice between waiting on interminably or continuing towards a definite goal. We decided to continue.

We felt, I far more than my men, a certain lowering of vitality, due to the very high altitude on which we stood. And lastly, without daring to put it into words, we realized only too well that, if our two friends were still on the pass, they were almost certain to be dead.

. . . . .

Then began an endless descent towards the low-lying valley of Dekho, of which we knew nothing except that there were men there who lived on other things besides plunder.

Although we were no longer delayed by the vagaries of our yaks, we nevertheless made slow progress. From time to time, in an agony of pain, poor Tchrachy would let himself slide off the horse's back, begging us to leave him behind. We pointed out to him that it would soon be dark, and that if we hadn't reached the valley before then, we should be unlikely to find help that night.

The track was at first quite easy to follow, but the lower we got, the more rugged it became. The valley changed bit by bit into a wild, narrow gorge. The fall of the river increased and its course widened. We had reached the forest, a real alpine forest of juniper trees and thick undergrowth, which multiplied the obstacles in our path. The entwined branches of dead or living trees, the stone debris and boulders compelled us to cross constantly from one bank to the other. Yong Rine had removed his boots and rolled up his trousers. Unfortunately I couldn't follow his example; my feet would soon have been cut and bruised and would have prevented me walking. So I had to wade through the water with my boots on, and when I got out they weighed me down like divers' boots soled with lead. It was an appalling effort lifting them off the ground.

We must have been advancing at the rate of a mile and a half an hour, and the sun had already disappeared behind the left slope of the valley. Anxiously we watched the shadows creep up the other side of the mountain. The gorge was plunged into semi-darkness which thickened every moment.

We kept turning round to see if anyone was pursuing us. Once we thought we saw something moving in the distance. This threat from behind was unbearable. It made us forget that there were dangers ahead as well as in the rear. But at nightfall an incident brought this fact home to us. We were trying to penetrate through a dark, narrow defile lined with thickets. Yong Rine was leading the way, dragging the horse by the bridle. Suddenly he came to a dead stop, and seized hold of the strap of his rifle. What was it now? I drew up with him, with my hand on the trigger of my revolver. Petrified, his eyes wide and staring, the Tibetan was searching the dark depths of the forest. And yet there was nothing moving. Then I noticed, at his feet, in the middle of the track, a stone wrenched out of the ground and turned upside down, with a few twigs arranged in the shape of a triangle. That was all. But in these completely virgin surroundings these faint indications of human life are none the less terrifying. In every nomad country there is a secret language of the road known only to initiates, herdsmen or bandits. The earth being still fresh, these twigs indicated the recent presence of some traveller or caravan, and in this hostile country we could only consider them a bad omen. Perhaps they marked the meeting-place of some sinister band plotting against us. For the first time I foresaw the possibility that our enemies had posted a second body at the far end of the valley to waylay those of us who had escaped the massacre.

However, we saw nothing. And having no choice of direction we continued our journey, our eyes on the alert, our fingers on the triggers of our guns. There is nothing more frightening than woodland for tracked men. I imagined Liotard still alive, suffering unspeakable torments in his complete solitude.

The valley slopes were so steep and the alpine forest so dense that it was quite impossible to advance in any direction except along the thalweg. Sometimes we had to wade through the stream itself. If anyone were waiting for us lower down they couldn't fail to see us. We had an impression of running the gauntlet of a million eyes, black eyes peering at us everywhere out of the thickets. So, when night fell, we felt relieved, in spite of the increased difficulty of making headway.

It was a bright moonlight night, but we could not see the moon, for it was already low and was about to set. Moreover, the gorge had become so deep that we were in complete darkness. Only now and then could we perceive, far, far above our heads, an occasional grass-grown ridge, bathed in a blue light. But when the moon had entirely disappeared the darkness became so thick that we felt we were pressing against some solid substance. Nevertheless, we pushed on desperately, intending at all costs to reach the valley junction. We were very often forced to retrace our steps to get back on to the faint track which Yong Rine followed by instinct, like an animal. We knocked against overhanging branches, rocks and fallen tree-trunks, and sank knee-deep into swamps. And always this swift, widening torrent roaring in our ears, which we had to keep crossing and re-crossing. I was completely blinded and held on to the horse's back, following the noise of crackling undergrowth and broken branches.

For the first time I realized, in spite of several months of travel, how much I had remained a townsman, a man deprived of certain primitive, elemental faculties, and at that moment I would have given all my cultural knowledge to be able to see in the dark, like a Tibetan peasant.

. . . . .

And still we continued our descent. . . . And the slope of the valley became steeper and steeper. . . . The stream threatened at any moment to become a waterfall and to disappear into an abyss. There was no sky to be seen through the thick overgrowth of trees so we felt as though we were going down a spiral staircase into some subterranean vault. The dark trench seemed to be leading down to the centre of the earth, perhaps even to the infernal regions. Could it possibly be taking us to a peaceful, inhabited country?

And yet there must be some end to this interminable descent. I was certain that we were not very far from the valley junction. I felt no symptoms of breathlessness; on the contrary, for some curious reason, I felt a renewal of vitality, in spite of the fatigue and the incredible trials of the journey. So the air must have become thicker, which meant that we had reached a very much lower altitude. I knew quite well that in this district there were no valleys lower than 11,500 feet, so we had probably descended between 4,000 and 4,250 feet. So we really were approaching the end of our nocturnal ramble. This thought was enough to make us see imaginary houses everywhere. But these komba hallucinations always disappeared like mirages in the desert.

As was to be expected, the approaches to the junction were almost impracticable. The track ascended, descended, running up against cliffs and twisting in and out through a thick forest of shrubs and pine-trees. In our pig-headed efforts to push along we wasted more than an hour, making no progress and running the risk of toppling over at any moment into the ravine. At about nine o'clock that night, discouraged by our futile efforts and exhausted by thirteen hours of tramping without food, we decided to stop where we were and wait for daylight.

That night was for me a gloomy vigil, which I had to endure in silence, knowing nothing of the fate of my companion.

We had lain down in the forest on a rain-soaked slope, slightly above the level of the river. Yong Rine, with his Tibetan flint, had once more accomplished the miracle of fire. It was really a miracle in this damp gorge to set light to a twig with the mere spark of a silex. This primitive tinder, shaped like a wallet, is similar, the ethnographers say, to those used by the Vikings. It spared us a great deal of suffering for we were soaked up to our thighs, and I do not know how we could have endured the night temperature without a fire. It was nearly mid-September, and at this altitude of nearly 13,000 feet the nights were already cold.

Now we felt an animal sense of well-being in curling up close to the fire, lying quite still and forgetting our cares by merely gazing at the flames. Ever since that unhappy evening I have sympathized with fire-worshippers. Yong Rine had built up the fire between the two roots of a juniper tree, and its sap soon fed the blaze. The leaping flames burned out a crevice in the tree-trunk and lit up the trees around us, so that our camping-ground must have been visible from a good distance. But our need for rest and warmth was so great that we did not worry about that.

As soon as the fire was high enough to allow us to see we examined Tchrachy, who was lying exhausted on the ground. It had been touch and go with him, for he had been struck in four different places. In addition to the scratch on his chin, he had a large blood-swelling on his thigh, caused by the ricochet of a bullet off his saddle, and another scratch on the calf of his left leg; the boot had been torn by the shot. None of these wounds were serious. But I couldn't make up my mind about the wound in his back, just above the kidneys. The bullet had entered through his left side, crossed over to the right, without making any deep incision, and had remained lodged under his skin. It must surely have touched the spinal column in its passage, but, feel the vertebræ as I might, I could not find anything abnormal apart from the agony which poor Tchrachy seemed to suffer whenever I touched him. I already began to suspect the explanation of this curious wound, which was later confirmed by the facts. The bullet had not had enough force behind it to break the bones, but had merely twisted one of the vertebræ. The bad quality of the powder had saved Tchrachy from a mortal wound, but it had nevertheless given him a terrible shock.

We could easily have removed the bullet by making an incision with a dagger, but we had nothing to make a bandage of, so I thought it would be foolish to open yet another wound in the man's back. I thought it better to reach Dekho first and perform the operation there.

My two companions in misfortune were now feeling slightly better, and soon recovered their spirits. This was in keeping with the Tibetan character, but nevertheless astonishing. They began discussing the details of the fight like two college boys after a football match. They didn't feel any shame at having been beaten, seeing that our enemies had outnumbered us and, moreover, had had the advantage of position. The death of our companions didn't worry them much either.

Did they feel any deep hatred for the brigands who had so viciously attacked us? I wasn't sure about that. It was clear that if they had held their ground, they would have paid them back in their own coin, with interest. But the method of the attack, surprising us like that in ambush, is so typically Tibetan that they didn't feel to the full its baseness. But this did not prevent them making plans for revenge, each more puerile than the last, but which would doubtless occupy them for the rest of their lives. For in Tibet vendettas persist longer than in any other country.

Their excitement was natural after such a terrible day, but it only served to increase my own overwhelming sense of loneliness. Full of animal joy at having escaped death, Yong and Tchrachy had forgotten their companions and my grief. Even the mention of Liotard was hardly enough to stop their flow of running commentary. Tchrachy, the more sensitive, or perhaps the more tactful of the two, possibly felt some genuine emotion and when I spoke of Liotard his fine eyes clouded over. But both men, though in no sense ruffians, were nevertheless of the stock of wild nomad tribes, who do not linger beside fallen companions, and who leave behind all who are too ill to keep pace with the caravans, with a provision of food and water. This is such a well-established custom that it seems quite natural to those who practise it.

Their indifference in regard to Liotard shocked me less than their scorn for the dead man who, when all's said and done, had been their tent-mate for three months. This scorn could not be attributed entirely to the cowardice shown by our unfortunate cook. There was something in it of that deep-rooted race-hatred, an ingrained bitterness of which they themselves were unaware. Poor old Tze! Humble participant in the tragedy, menial in a perilous expedition, he had never imagined himself dying a hero's death. And yet, without knowing it, he had given his life for the advancement of human knowledge. I couldn't think of him without pity.

In this deserted spot, in the reflection of these vaulting flames, we must have appeared a strange group. It only needed a touch of witchcraft to complete the fantasy. My companions did their best to increase the illusion. Unrolling the long, multi-coloured ribbon which he used to fasten the top of his linen boot under his knee, Yong Rine, rapt and solemn, kept knotting and unknotting it at a great rate and examining the varying combinations. This was his method of consulting the fates. Obviously he was trying to find out what had happened to Liotard.

The oracle's replies were not very conclusive. First combination: death, second combination: life. Yong chose the second: life. Lieu Sien-cheng had obviously gone back across the pass and we should find him again in some encampment. How far did he believe this interpretation? I fancied that a certain delicacy of feeling was prompting him to alleviate my distress.

But far from calming me, these experiments only agitated me more. I imagined our enemies, now that the excitement of the battle had abated, resorting to the same procedures in some equally sinister spot, performing strange rites to consult the oracles, winning over all the good spirits to their side, the spirits of the mountains, woods and rivers, and invoking the gods of the absurd Tibetan pantheon to bring about my death. It is difficult not to become god-infected in this country of Tibet, which possesses more deities than all the rest of the world put together. I, a solitary European, lost in this hostile, terrifying wilderness, fancied I saw in the magic-lantern colourings of the fire the reflections of those cruel and atrocious deities whom I had gazed upon in the temples, and I could not bring myself to believe in the benevolence of their intentions.

The situation was an exceptional one, so Tchrachy in my presence actually produced a delicately-modelled statuette from the big turquoise-studded reliquary which he wore slung across his shoulder. He kept turning this frail effigy round and round in his hand, gazing at it with a mournful expression on his face. Padma Sambhava, King of the Magicians, wrapped in orange muslin and faded paper inscribed with prayers, had, I imagine, failed in his mission, which was to protect Tchrachy from the dangers of the road. And yet Tchrachy kept repeating to me that he had paid sixty good rupees for his gaou. He strongly suspected the lama from whom he had brought it of having swindled him. What a difference, he said, from the charms which he had carried with him during one of his previous journeys, and which had brought him back safe and sound to Tatsienlou after the most thrilling adventures! The fact that he had got them cheaper than this one obviously added to their merit. This mixture of shrewdness and superstition is one of the most astonishing characteristics of the oriental mind.

I pointed out to Tchrachy that his lama had not treated him so badly after all, because he might think himself lucky not to have been killed by one of the four bullets which had struck him. At this Tchrachy burst into peals of laughter.

Both uncle and nephew were full of admiration for the way in which I had escaped from the bullets. I could not see that there was any merit in it, but they congratulated me with upward jerkings of the thumb. I had discovered that as well as the tear in my jacket another bullet had torn my trouser-leg. Yong Rine and his nephew kept examining the holes, which testified to my good luck, and I could see that for two pins they would have asked to see my talismans. I assured them that, with fate on my side, I owed my life, as they did, to the bad quality of our enemy's ammunition. Upon which my companions began imitating the noise of popping guns and whistling bullets with such accuracy that I begged them to desist and take a little rest.

In spite of my exhaustion I felt neither sleepy nor hungry. But on the other hand my throat was parched with thirst. Yong Rine had already gone down several times to the stream and brought me water in my cap which was the only container we possessed.

It would be difficult to describe my feelings during that long and painful night. I was still too agitated in mind to realize the full sadness of Liotard's loss, and as is always the case on such occasions, my mind alternated between wild hope and deep despair. I kept repeating idiotically to myself that a man who had been my tent-mate for the last five years could not have disappeared suddenly like that without any warning. Lying with my feet to the fire, I looked enviously at Tchrachy and Yong Rine; Tchrachy kept turning and twisting about, but, in spite of his wound, managed at last to drop off to sleep; Yong was sleeping peacefully. At a little distance our poor horse, tethered to a tree, was gnawing the branches to stave off his hunger.

Oh, that long, long night spent waiting for the dawn and calculating the chances of seeing Liotard alive!

. . . . .

At last a misty dawn appeared. My two hirelings rubbed their eyes and stretched themselves. They had lost their last night's fluency and their voices were hoarse. Their faces bore marks of twenty-four hours of exhaustion. I too was very weary, but I did not worry for I believed we were near our goal. We started off again slowly, without a word, on empty stomachs.

After the first hundred yards I was glad not to have persisted on the previous night. In the darkness we could never have followed this path without breaking our necks. It had become so shapeless that it was like a giant's causeway, leading us over rocks and boulders and sometimes overhanging the stream. Tchrachy was forced to dismount and continue on foot. The wretched fellow limped along painfully, his back bent double, his legs straddled, without a word of complaint.

As I had imagined, we were not far from the junction. Suddenly, after surmounting a little rise in the ground, we saw a wide open space in front of us. At last we had reached the other valley.

It was certainly no spacious valley, this narrow, deeply-embanked ditch with slopes encroaching upon the terraces which here and there bordered the river. But we had been clambering and slithering along the bottoms of so many gorges that by contrast it appeared vast. We were still about a hundred yards above the river's edge, and the water itself was barely visible because, although the weather was fine, banks of mist rose up here and there like white translucent walls. For this reason I could not determine the direction of its flow. Were we on the right or the left bank? I must in truth confess that this matter of professional interest did not hold my attention long. What really held my gaze in this new landscape were the four or five herdsmen's tents pitched on the opposite bank, and the little black dots which encircled them. A camp at last! I felt the joy of a drowning man who sees the life-boat rising over a billow.

Nevertheless we were not yet at the end of our sufferings. But the sight of our fellow-creatures had revived us, and we clambered down towards the river's edge almost at a run. But an unpleasant surprise was awaiting us. We were on the right bank of the stream, a regular little rolling mountain torrent, more than a hundred yards wide, hemmed in on both sides, swift and choppy with waves, which we could see at a glance was absolutely unfordable. Although we were within hailing distance of the men who were peacefully intent upon their business, an ocean could not have more effectively separated us. As far as the eye could see in this twisting valley, there was no indication of the presence of human beings on our bank. Not a blackened stone, not even a broken branch. The forest which overlay the steep slopes was obviously virgin.

In my disappointment I almost forgot that I had just reached one of the principal objectives of our expedition, the upper course of the Tong. Moreover, when I did realize it, I felt no sense of elation. What good was it to have discovered that the source of the river lies at the foot of the glaciers of the Bayen Khara mountains, seeing that I had no instruments, not even a compass, to establish our position or to take bearings, and that all our previous notes had been lost? Ever since last night I had avoided thinking of the results we had achieved during our long weeks of travel, about Liotard's and my work, all of which had been stolen from us and could never be recovered. I began to feel a superstitious fear of some malevolent force whose aim was to preserve from our impious prying eyes the last unknown territory of Central Asia, the high table-lands which have been the haunt of barbarians since the early ages of mankind. And as I looked down at this rolling river which I christened Tong, I could not help feeling that probably neither Liotard nor myself would ever mark it on a map. And in my exhausted condition I came near to despair.

It was useless to remain there gazing at the pastoral scene in front of us. We must at all costs resume our journey. But in what direction? Clearly this was the river which the people of Chortaintong had called the Ngolo Ma Tchou.[1] Therefore Dekho must lie somewhere farther downstream. We had better continue down the valley.

And so we pursued our weary course. Our resistance was weakened, more perhaps by the disappointment we had just suffered than by exhaustion and lack of food. Although for the last twenty-four hours I had swallowed nothing but icy water from the mountain stream, I still did not feel very hungry. We walked on with slow, dragging steps and at about eight o'clock reached a point where a large tributary joined the main stream. The mist had thickened to such a point that it was impossible to find a place through which to wade. Yong Rine suggested crossing the stream on horseback in two journeys, two of us seated on the horse's back. It would be a dangerous experiment. The horse was obviously as tired as we were and with two men on his back would as likely as not collapse in midstream. It would then be all up for the riders, and one of us would be left alone. The idea that I might be the one left on the bank

[1] In Tibet, as almost everywhere in Asia, the rivers have local names which are the only names known to the inhabitants of the districts which they cross. No one in Tibet knows the name Yang Tze, although the river has its source there.

frightened me more than the prospect of being carried away by the flood. No. That plan was definitely out of the question.

Moreover, we were at the end of our tether. Tchrachy, whose sufferings had eased in the night, was now again in great pain; the efforts which he had to make to break through difficult places irritated his wound terribly. How much farther should we have to walk before reaching Dekho? Should we get there at all with our single horse? I felt my strength already failing and doubted my power to continue many hours longer. We should have to think out some other plan.

We soon came to a decision. Yong Rine was to take the horse and ride along down the valley until he came to a settlement where he could find two horses and some food. If he found Dekho, the lamas of the monastery would surely come to his aid. He could spur on our poor horse to a slightly swifter speed, and would reach Dekho much quicker by himself than in the company of a cripple and an exhausted European. So we exchanged weapons; I gave him my revolver which was easier to hide, and we kept the rifle.

Yong immediately leapt on to the horse and drove him straight into the stream. The animal sank breast-high into the flood, and stepped gingerly forward, feeling the bed of the river before leaning his whole weight on it. His rider merely held his head above water, leaving him to choose his own course. Finally they scrambled up the opposite bank and we saw them disappear into the fog.

. . . . .

Then began for Tchrachy and myself a wait which turned out to be much longer than we had expected. Hours were to go by, endless hours of increasing anxiety.

At first we were glad to be able to relax and not to have to worry about our safety. We lay down quietly and allowed our minds to wander. After the enormous physical effort of these last twenty-four hours, without any interval of food or sleep, I felt a sort of psychic weakness come over me. My head seemed to be even emptier than my stomach, and I did not suffer terribly at the thought of Liotard.

When the morning mists had finally dispersed a beautiful day dawned. Tchrachy, like a good Tibetan, immediately stripped himself to the waist. Seated on a little green mound still festooned with flowers, we must have resembled from a distance two Parisians taking a sun-bath in some secluded suburb. I don't know why this strange simile should suddenly have flashed across my mind, recalling in every detail picnics in the forest of St. Germain. Perhaps because ordinarily we should have been surrounded by tins of preserved fruit and tissue-paper; perhaps because grass is the same everywhere. Whatever the reason, the most extraordinary associations of ideas floated at will through my anæmic brain. We had nothing whatever to do but wait quietly for our friend's return, so I allowed my mind to wander in endless ecstatic fantasies, all tinged with a slightly suburban sentimentality, in vague longings to mingle once again with creatures of my kind, people who spoke my own language.

Several hours passed in this way, without our exchanging a word. There was still no sign of Yong Rine, and the sun was already setting. When it reached the mountain-top darkness and misery mantled us like a cloak. We began to wonder whether Yong Rine had met with some disaster.

Were we going to expire quite stupidly of exhaustion and hunger, with tribesmen passing to and fro on the opposite side of the stream? The situation was really too ridiculous.

Without knowing exactly why, we started watching the people on the left bank. Late in the afternoon our cries and wavings attracted the attention of two lamas. They came and stood for a while at the water's edge, looking at us in amazement, and then turned away and continued their journey. The rippling of the waves had drowned our voices. That was conclusive; we had only ourselves to rely on to get us out of our trouble. There was still no sign of Yong Rine, so there was nothing for it but to get up and start off then and there. To-morrow it might be too late; there was a risk that Tchrachy might not be able to move, and if the weather let us down our position would soon become desperate. From now on I had no more illusions. I knew that Tchrachy could not walk very far with his wound, but I counted on this move reducing the fear of death which was gradually taking possession of us, a fear more unbearable than yesterday's terrors, because now the danger was not immediate.

Then something happened which put an end to our doubts and at the same time broke the monotony of this dreadful day. After the mist had lifted we had located a spot where the stream was fordable. We were just stepping into the water when two riders appeared far below us in the valley. They were on our bank, and they were coming towards us. I thought at once that it was Yong Rine accompanied by a native, and I gave a sudden shout of joy which made Tchrachy jump. But Tchrachy, more sceptical, did not share my excitement, because it was still impossible to distinguish the two men. Alas, the nearer they approached the more our hopes sank; the two riders were strangers, Ngolos of some kind. What were they doing here, and where could they be going at this late hour in the afternoon, on this lonely path, without any baggage on their horses?

Tchrachy watched them in silence for a while, and then uttered the word which I was dreading to hear: "Tchapa."

I don't think there is any other Tibetan word so familiar to the traveller. I know that in Tibet all strangers are suspect, but I am sure they have not all the classical "comic-opera brigand" appearance of the two who had just broken upon our solitude. It is almost incredible that there is still a corner of the world where one can meet such sinister individuals.

They were scantily clad in inverted sheep-skins, and their bared right shoulders revealed that they had no undervests on, or trousers either. Their long, naked feet protruded through short stirrups. This get-up, which would have appeared wretched in the meanest herdsmen's settlements, contrasted strangely with the magnificence of their mounts. Horses like that could belong only to chiefs or felons. Moreover, if we harboured the smallest doubt about the nature of their business, the sight of their guns was enough to dissipate it. They weren't forked guns; they were definitely weapons of Chinese or European manufacture; useful equipment for professional brigands.

We were already so desperate that this fresh blow did not hurt us at all. It did not even surprise us. But I wondered idly whether these two highwaymen formed part of the troop who had attacked us and had deployed from the others to complete their work of destruction and polish off what remained of the expedition. For the moment that was the only thing that mattered. If we had to deal with old friends who were on the look-out for us, the danger was infinitely greater than if this were a chance encounter. No Tibetan ever attacks on the impulse of the moment. They only decide to attack after long palavers which often last several days.

The surprise on the faces of the newcomers when they got to within a hundred yards and could examine us reassured me at once. They reined in their horses. The appearance of a strange monster with a red beard in this lonely spot must have given rise to dark thoughts in their primitive minds. They were obviously seeing us for the first time.

They remained looking at us in silence for a while; then, without a movement of self-defence, turned and spurred their horses into the brushwood. We tried hard to track down their route, but no further sound betrayed their presence. Probably they had ambushed themselves in the undergrowth and were waiting. What for I did not know. The sensation of being devoured by the eyes of these hidden ruffians was an added strain on our tired nerves. I really believe that in the long moments which followed the whistle of a bullet would have relieved our tension. We knew that our ammunition would not last longer than five minutes, but we had reached the limit of our endurance.

Thank God this intolerable situation did not continue very

long. That day, the longest and worst of my life, was drawing to a close.

Soon two other riders appeared.

This time there was no doubt. It was indeed Yong Rine returning with a big, gawky, boisterous-looking devil of a man. When I saw him I thought at once that our new friend looked just as suspect as the rogues lurking a few yards away; in Tibet it is difficult to distinguish a robber from an honest man. It is true as, Père Gore has remarked,[1] that these professions are interchangeable. In any case the new addition to our party did not appear to lack courage.

On seeing me he burst into a great guffaw of laughter. Then, spurring his horse on to a rise in the ground, he turned and faced the two bandits, ostentatiously placing his rifle on his thighs.[2] He had at once sensed the danger of our position, and wanted to show these highwaymen that they had a determined fellow to deal with.

Yong Rine was in the highest spirits. He was quite restored in health and seemed delighted with the results of his expedition. With repeated upward jerkings of the thumb he told us how he had reached Dekho, where he had been very well received by the Grand Lama, who had immediately presented him with our new companion and two yaks. Whereupon he had at once returned to find us.

Squatting on the river bank, I quickly swallowed a mouthful of tsampa. But there was no time to lose. The Ngolo was keen to leave this unhealthy district and did not like the idea of being caught in the dark. We were just as anxious to start as he was.

I mounted our unfortunate horse, now freshly saddled, and my two men mounted the saddled yaks.

[1] Notes on the Tibetan frontier-lands of the Sseu-chouan and of the Yunnan (*Bulletin de l'Ecole Française d'l'Extrême Orient*, 1923).

[2] The Tibetans wear their stirrups very high. Thy literally squat on their horses in an attitude something like jockeys. In this way they can rest their rifles horizontally across their thighs. This gesture indicates that the man is on the defensive. Ordinarily the gun is carried slung over the shoulder.

Tchrachy, realizing that he was safe once more, regained his high spirits. He lay spread-eagled on his yak, and his physical sufferings did not prevent him shouting with laughter at his uncle's very unwarlike appearance. A man astride a yak is always slightly ridiculous, especially if the animal keeps trying to throw his rider by the most underhand methods.

. . . . .

Finally, after a day which seemed to have lasted a hundred hours, darkness began to fall and through the evening mist I distinguished a few stone houses shaped like square towers and the horned roofs of a temple. Some men were advancing towards us, barefooted and squelching through swamps; it had apparently been raining. In the light of the wood torch which they carried aloft I could see shaven skulls and frightened eyes. These people, disturbed in their solitude, betrayed their anxiety by hoarse cries, and I could see sword blades and rifle barrels shining in the darkness. Dogs, startled by the noise, began howling and barking.

As I entered this haven I scented suspicion, fear, and mutual mistrust.

CHAPTER 10

THE COMPASSIONATE LAMAS

For modes of faith let graceless zealots fight;
His can't be wrong whose life is in the right:
In faith and hope the world will disagree,
But all mankind's concern is charity.

A. Pope. Essay on Man.

THE noise of crackling wood, the light of the first flames which shines pink through my closed eyelids and the smell of burning resin tells me that my host is preparing my breakfast. I can sense him crouching close behind my head, turning his eyes towards me and then back to the fire, watching for me to wake. I lie there motionless, prolonging as far as possible that half-state between waking and sleeping, frightened at the thought of suddenly returning to reality, postponing the inevitable moment when I shall have to respond to the smiles of this lama whose guest I am.

For the moment I wish to remain a sleeper, for sleep brings the blessing of forgetfulness to king and cottager alike. I delay the fatal moment when I shall have to fix my eyes on the creatures and objects which surround me and with whom I seem to have lived for years.

Sooner or later I shall have to open my eyes, sit upright on the filthy blankets which compose my bed, shove back the wooden saddle which serves as my pillow, and try my utmost to understand the solicitous inquiries of this over-attentive lama. Then, when he's finished, it will be his mother's turn. She'll come in, half-dressed as usual, her naked bust protruding above a yak-skin coat rolled down to her waist, looking like a half-skinned rabbit. She, too, will want to console me, and to force down my throat cup after cup of salted and buttered tea and a bowl of tsampa. She irritates me slightly, this old Tibetan momma, and yet I like her and her presence has sometimes a soothing effect on my heart hardened by my experiences. She is so good-natured that I can no longer think her ugly in spite of her flopping breasts and cropped skull. For she, too, is a lama, and therefore scrupulously shaved.

This monk's cell in which I am lodged stands at the foot of the monastery, on the fringes of the forest whose lowest-lying trees are only fifty feet from the door. It is my fourth awakening, or rather the first time I have not pretended to awake. For the first three nights after my arrival I don't think I closed my eyes longer than ten minutes, so great was my anxiety. I did not yet know for certain what had happened to Liotard, and I still clung to an excruciating hope which almost amounted to madness.

But now all is over. My wild hopes are dead. It was yesterday that I learnt I had nothing more to expect, and that I should never see Liotard again. And, on the very night of the return of the rescue-party and of the confirmation of my most hideous forebodings, some perverse physiological process caused me to fall into a deep sleep.

The morning after our arrival the Grand Lama had sent a small body of riders, priests and laymen, to accompany Yong Rine to the fateful pass and search everywhere for my friend, dead or alive. I was too exhausted to go with them. The efforts which I had made in these last two days, deprived of food and sleep, would have been a strain at any altitude, but, on these considerable heights, they had completely overtaxed my strength. I would have been physically incapable of returning to a height of over 16,000 feet, nearly a mile of which would have been spent climbing, and the greater part of which had to be done on foot up a gorge. Above a certain altitude the lowlander becomes a cripple. Even had I wanted to go the Ngolos wouldn't have welcomed me. My presence in the rescue-party would have increased the dangers of the road. Already a rumour had started —God knows how news travels through these mountains—that the tribesmen were still intent on killing me, and that sinister figures had been seen on the mountain paths which surrounded the monastery. For that reason the whole party, priests and laymen, had gone off armed with guns and swords, and bristling with ammunition pouches.

So I had remained behind with Tchrachy, who was himself prostrated and suffering greatly from his wound. Then began for him and me another interminable wait, this time indoors.

The Tibetans told us on leaving us that they would be back the following day. All that day I waited, pacing up and down between our house and the entrance to the village, my heart pounding in my chest, heedless of the curiosity of the monks, who stood watching me for hours on end, amazed at my state of agitation and at seeing me expending so much useless effort. When the sun finally disappeared behind the mountain I thought I should go mad. Having eaten and rested my normal reactions had returned, so that at the thought of my lost friend I began to suffer horribly. I found myself talking to him, begging him to return, uttering things which in our rough man-to-man relationship I would never have thought of uttering. I still felt such sick sorrow, such an agony of misery that there was not yet room in my mind for the selfish fear of being left alone, defenceless and friendless, in the centre of Asia. I wished, I longed, as never before in my life, for one thing only, the return of Louis.

It was three days before the Tibetans returned. I was in the house, mechanically poking the brazier on the hearth when I heard the sound of horses' hoofs squelching through the mud. I had watched for them ceaselessly for the last two days, and now they had surprised me indoors. Tchrachy and I sprang to our feet and, very tense, stepped towards the door. Yong slipped heavily off his horse's back, his face drawn and white. He came in without looking at me, and threw Tchrachy's hat into a corner. He had found it at the scene of the attack. Then he went in silence and squatted down by the fireside. His friends had followed him into the house exhausted, with their guns in their hands. The woman lama and her son were also present. There they stood all of them assembled, in the smoky atmosphere of the room, acquainting me by their silence of the extinction of my hopes.

. . . . .

I shall not try to describe the misery which I endured from now on, those long days spent in the smoky atmosphere of that wretched, windowless room with its soot-varnished ceiling. Despair is not a subject to write about. But I can describe the terror which seized me at the thought of being left alone. Everything around me was so alien to me and to what had previously been the background of my life that I began to lose the sense of my own personality. I was becoming a stranger to myself, and, by a curious process of dissociation, looked with pity and astonishment at the wild, bearded man begrimed with dirt, seated in a lama's dwelling, wearing an expression of pain on his face, who yet bore some vague resemblance to myself. And as though it concerned some stranger, I began calculating his chances of escaping alive and one day returning to his own country. That country seemed so far and so unattainable that an obstinate doubt wormed its way not into my own soul, but into the nerves and marrow of that stranger, imprisoned in this immense Asia. For he was, I was, in the very heart of this vast continent which seemed to form around him and me concentric circles of enormous mountains, gigantic rivers, stretches of country either deserted or over-populated by various races all at war with one another, circles which bound us like Dante's Inferno, and which we should have to break through in order to reach the sea which had brought us here and by which we still hoped to escape. The sea, which meant freedom, life and joy to us!

I also felt an intense longing to escape from the dirt of my surroundings, from the filthy yak-skins which I was compelled to wear, from the verminous insects with which my companions had infected me. I was becoming more and more like the Tibetans with whom I lived. At night I would dream of polished floors, of light paint, of well-planted avenues of trees, of trim, beautiful girls. There was so much ecstasy in these dreams of mine, that when I allowed myself to drowse I actually forgot that Louis was dead. Then I would wake with a start, expecting at any moment to see him enter the room, this dark, gloomy room where I was forced by the constant rain to spend these endless days.

Thc fine weather had only lasted three days. What we had taken for autumn was only a truce. Soon the weather changed for the worse, and clouds closed down over the valley like a lid over a coffin, increasing the oppressiveness of the atmosphere in which I lived. Could one really call it living? I had no book to read, no paper to write on, and I did not know enough Tibetan to be able to converse with the good people who sheltered me. A prisoner in solitary confinement can exchange a few words each day with the warder. But even that comfort was denied me. My loneliness was complete.

Nevertheless, the good people exhausted themselves in efforts to distract my mind and to drag me out of my state of isolation, for they seemed to appreciate my tragedy. It is true that their pity had not been spontaneous. When I first arrived no one dared take me under his roof. I was a stranger from a far country which they had never heard of, hardly a human being to these religious folk. Moreover, I was an incarnation of misfortune, and who knew what curse I might bring upon them! So they had deposited me out of doors under the porch of a building in the courtyard, with a few blankets thrown to me out of charity.

I had sunk down on these blankets in a state of utter exhaustion. But a few minutes later, gasping and shivering and swallowing mouthfuls of fog, I had got up and gone into the common room, where the lamas were listening to our adventures as related by my jabbering servants.

When I appeared there was a general silence. They all turned and stared at my face which was reflected in the firelight.

It was the first time they had set eyes on me. All of them seemed to be waiting for me to reveal the mystery which I carried within me. But I merely walked up and joined their semicircle round the fire, reaching out my hands to its warm blaze as any traveller would have done. Probably I had also smiled at them, and my smile broke the magic spell. At all events they suddenly lost their awestruck mien, and rushed at me in a body, smiling at me, tumbling over each other in an effort to please me, to give me proof of their goodwill. They sat me by the fire, wrapped me in skins and fed me, and then insisted that I should lie down, and I believe actually tucked me up. I had become a man like any other, a wayfarer who had a right to a place by the fire. It was then that I realized the goodness of these simple folk, whose pity had been roused by a simple smile.

. . . . .

Tchrachy still had his bullet lodged in his right side, ever since its extraordinary course between the skin and the vertebral column. The morning after our arrival I suggested removing it for him; everyone acquiesced with true oriental politeness, but no one did anything to help me. It appears that I know nothing about it. A surgical operation is something quite different from what I had imagined; it has to be performed with a paraphernalia of rites which I had no idea of. I only discovered that to-day.

They came, a dozen of them, to assist the lamasery doctor who is to remove the bullet. The operation itself takes only a few seconds, but the preliminaries are endless. It is not a question of disinfecting the instruments or the wound itself; here, prayers and observances take the place of those precautionary measures which our Western materialism insists upon, and it would never occur to a Tibetan to cleanse a sore. On the contrary, his main object is to contaminate it with repulsive unguents, whose concoction is entirely subjected to the law of fantasy. There is a dash of everything in them, from aconite to Chinese ink, from musk to bear's gall and peacock's feathers.

The patient is laid out on planks, while the lamas come and squat round him in a circle on the ground, with their large wooden prayer-books resting on their knees. The good people are full of gleeful anticipation of the coming ceremony, presumably because it breaks the monotony of the daily monastic routine. The weather, for a wonder, is fine, so the scene takes place out of doors, under the porch of our house, facing the forest. And there is no lack of gaiety about it. This solemn medical ceremony is stamped with the usual Tibetan good-humour. The lamas break in now and then on their ritualistic chantings to drink an informal cup of tea or to crack jokes with the patient, who, for his part, waits calmly until the ambient air is favourable for him to take down his breeches and offer his loins to the artist's lancet.

The latter seems rather pleased at this opportunity of showing his skill. He has always been the doctor of this lost monastery, so he has obviously not attended the medical lectures at the college of Mont Chakpori, on the outskirts of Lhassa, which, under the tutelary protection of the Eight Buddha Healers, is the training centre for the practitioners of the great lamaseries. He can only have developed his science through tradition, handed down orally from generation to generation by the incumbents of his profession. But in Tibet, more than elsewhere, medical practice is tinged with wizardry, so there is something quackish in his method of attack. The operation is quite a simple one, for the bullet in Tchrachy's back forms a small, well-defined lump under the skin, and a neat incision would be enough to loosen it. Nevertheless the doctor brings with him his whole outfit, a sort of linen satchel, containing iron implements, a scalpel, tweezers, lancets, all more or less similar to those of a surgeon's emergency kit, the whole wrapped in filthy pieces of gauze which the lama fingers quite happily under the approving eyes of his confrères and of his future victim, who, for his part, appears entirely unconcerned.

I feel cold shudders go down my spine when this curious doctor starts solemnly and conscientiously wiping his lancet on a carefully-folded compress which bears the sordid marks of blood and pus from the abscesses of his former patients, collected over a period of years. But I take care not to interfere, realizing that they would not understand, and that they might even take my intrusion for an act of malevolence towards my hireling, or read in it scorn of the religious observances which are part of the operation, and which, in the minds of these people, are the basis of the whole affair.

The vehemence of the praying increases, and the lamas strain their necks to get a full view of the spectacle. The man bends over the patient's back, mumbling a magic formula. A thrust of the blade with a sure hand, and out jumps the bullet! For Tchrachy it will be one of his most precious talismans. There's nothing to be done now but staunch the blood with that hideous compress and spread an ointment over the two gaping holes on the patient's back. The magic of this ointment must indeed be strong to prevent this curiously tended wound from becoming infected.

The lamas now congratulate Tchrachy at having been so happily delivered of his troublesome piece of metal, and then depart chattering together, delighted to have fulfilled their duty, and quite unconcerned as to the ultimate result. Tchrachy, who has endured the whole process with great courage, comes and lies down by the fire, giving us a wry smile. There will be some supuration of the wound to-morrow, but Yong Rine will come and spit on his back and rub him, and in a week's time it will be gone. Nature works wonders, and the Tibetans are tough. Who can say, after all, how the thing would have turned out if I had interfered?

. . . . .

Thus passed the hours and the days, interrupted now and then by curious little incidents. We did not know what was to be the end of it all. In the first place we had had to wait until Tchrachy was in a fit state to travel. But now that at a pinch he could ride, we had to plan our departure—let us rather say escape—because the first day's vague alarms were becoming more manifest and we knew for certain that one or more bands of riders were wandering at large in the surrounding mountains. From time to time strangers would appear in the monastery, prompted perhaps by sheer curiosity, and would make straight for our house. The woman lama would rush and close the fence-work gate of the courtyard and stand facing the strangers, watching their movements, ready to give the alarm at the smallest sign of danger. The strangers would remain there watching her with surprise, taking good care not to touch the slings of their rifles or the hilts of their sabres, apparently only concerned to catch a glimpse of my face in the darkened room where their curiosity had forced me to take refuge.

Naturally this state of affairs gave rise to much talk. Monastic life, when it doesn't enforce complete silence, encourages the natural tendency of lazy folk to ceaseless gossip. Our intrusion brought a spice of novelty into the monotonous life of these lamas who, for the most part, had never ventured even as far as the market-place of Luho, that tiny hamlet which is for them a "town". So our adventures, our perilous situation, the dangers which lay ahead of us were the cause of ceaseless tittle-tattle everywhere, in the lamas' private dwellings, under the portico of the temple, or, whenever the weather allowed them to bask in the sunshine, in the courtyard beside the great chortain. Here, squatting or standing around in groups, dressed like Cæsars in purple togas, they composed a picture which any artist would have rejoiced in. The outcome of all these babblings was a great confusion, whose reverberations came to our ears in the form of contradictory opinions, strange counsels and confidences whispered in my men's ears. The whole monastic tribe was astir and plotting, and I thought to myself that if the bandits who were after us had any spies in the place they must be growing more and more perplexed about our real intentions.

The whole affair caused a certain amount of nervous tension, and the high dignitaries who were responsible for the safety of the community began to grow alarmed. The monastery was by no means powerful, and was an easy prey to attack. The population all told, including about twenty peasants who inhabited the six or seven farms in the neighbourhood, amounted to roughly a hundred, but its armament consisted only of a score or so of old-fashioned guns. The lamas, therefore, in spite of their kindly disposition, were obviously eager for us to leave lest we should bring misfortune on the whole community.

The Grand Lama had taken the matter personally in hand. He had graciously and with a due respect for the formalities returned the visit which we had made him in his house; we had found him there squatting Indian-fashion on an old carpet, in a wretched-looking room reached by a ladder at the far end of a stable-yard.

He was a tall, gaunt old man, very erect, rather dirty and wrinkled like a fig. I was astonished to find such natural dignity in this Father Superior. Without any feigned humility, he came and seated himself beside us in a friendly manner and began discussing the best means of helping us without compromising the interests of which he was in charge. Tchrachy, glib of tongue and literate, began expressing in beautifully polished and rehearsed language his affectionate esteem for the great man, who listened attentively, fingering his long rosary in his delicate hands, apparently flattered. My young Tibetan never forgot to remind all the lamas of his acquaintance that he had a brother who had won ecclesiastic honours in Lhassa and had become a high official in some big lamasery.

This state of affairs could not go on. We had to leave. I had first considered following the Tong valley, hoping thus to avoid crossing the plateaux and the passes. But I discovered upon inquiry that the valley became impracticable a few miles downstream from Dekho. So we then decided to double back on Serba, the last Chinese outpost we had met on our journey out. Of course, there was no question of taking exactly the same road back. For a moment I had toyed with the idea, in an uprush of wild longing to return to the tragic spot where lay the bodies of Liotard and Tze, in a rock-tomb erected by Yong

Rine and the lamas. But I had to renounce this plan, for I realized that it would only be courting fresh disaster. If it were true that our attackers, or other scoundrels of the same type (for we did not exactly know whom we were up against), were preparing to attack us for the second time, the most important thing was to put them off the scent by following, over this maze of mountains and valleys, the most unlikely, unfrequented and difficult paths. On this point everyone was agreed. The only thing to decide was whether we should go alone, with vague instructions as to the direction we must take, more likely than not to lose ourselves, or whether we should take as guides Ngolos appointed by the lamas. We would engage in long discussions behind closed doors, which these orientals seemed to revel in, and spend our endless days disputing this important point without ever reaching a decision. God, how the Tibetans love arguing! I was relying on the little bar of gold which I carried inside my belt to put a final end to their doubts.

All this time the rain kept falling and falling, a fine, drizzling penetrating rain which drowned the countryside and transformed this valley-bottom into a marsh. And at regular intervals I would hear the faint, mournful, obsessing call of the shell-trumpets, a sound which for the last few months had marked all the stages of my life and which still sent a shudder down my spine.

At the sound of the trumpets the lamas would rise obediently and proceed in single file over the light cantilever bridge which straddled the angry, roaring river at a point artificially narrowed to lessen the span. I would see them passing under the large, tattered black and white drapery of the porch which flapped in the breeze, and disappear into the cosy darkness of the temple. The evening service, held between five and six o'clock, was the most frequented of the ceremonies; it left the whole village empty, apart from the houses in which the few female members of the population lived.

One day, at an hour when the settlement appeared deserted, an incident occurred which was to bring the atmosphere of tension to a climax. The weather had suddenly cleared, and through a rift in the ceiling of clouds a weak sun had appeared, restoring light and shade to the general grey of the landscape and casting its rays here and there against the copper ornaments of the temple. Suddenly strident cries broke the silence, the cries of a female voice. Someone, probably my hostess herself, the woman lama who was praying away on the flat roof of the house, had just given the alarm signal to the monastery.

I had a vague feeling that the business concerned me, so I walked to the door with my two men, my hand on the butt of my revolver, with its poor complement of four cartridges.

A wave of terror seemed to have swept over the village. People were rushing out of their houses with guns and sabres, running this way and that, without quite knowing what they were doing. On the far bank of the river four or five men on horseback were jostling each other in an attempt to cross the narrow wooden bridge, while on the right bank where our house lay, men were urging them on with shouts and cries. They very soon vanished into the forest and I heard two shots ring out.

I still did not understand what was happening, and looked in amazement at Yong Rine who, rifle in hand, was himself rushing towards the forest. Tchrachy, still disabled, had remained at my side, and I tried to read in his agonized expression the reason for the excitement.

In the meantime the woman lama had descended from the roof. She was still stripped to the waist. Her face was contorted, and she looked at me wide-eyed, with an extraordinary expression of mingled affection and rage. Forgetting that I could not understand her, she jerked out incoherent phrases at me, half furious, half pitying. And all the time she kept trying to push me back into the dark depths of the house.

I resisted her, for I was determined at all costs to find out what fresh disaster had befallen me. The extraordinary creature then seized me by the wrist and, dragging me a few steps forward, pointed at the spot in front of the house where, a few minutes before, our horse had been quietly grazing. There was nothing now but a leather thong fixed to a peg, trailing in the grass. The horse had disappeared. Some man, probably ambushed in the forest, had waited for the evening service and then leapt out, cut the leading-string, mounted the horse and made off with the only means of transport which remained to the caravan. The good woman had been a helpless witness of the scene, and was now in a state of frantic rage, mingled with sorrow and pity for my misfortune.

This fresh blow of fate left me stunned. Once more the vague threats, the latent dangers were materializing, taking human shape and growing out of all proportion, removing our chances of escaping alive from this adventure.

But I really believe that this strange nun with the pendulous breasts and shaven skull was more affected than I by the disaster, and that if in her fury she had caught the thief she would have torn his face, gouged out his eyes and bitten him through the heart. She rushed to the little chapel in her house, up to the private altar, and returned trembling and transfigured, holding in her clenched right hand a copper bowl containing holy water. Then, buttressed on her straddled thighs as though to brace herself for a supreme effort, her body and head thrown backwards, her right arm outstretched with all its sinews taut, she yelled out into the air, in the direction where the thief had disappeared, the curse intended to nail him to the spot, with invocations to the genii, spirits and demons of the mountains, forests, streams and gorges to unite together in a hurricane. Then, with these curses on her lips, she flung out the contents of the copper bowl towards the horizon in a thrice-repeated frenzied gesture, while I, for my part, clutched hold of the door-frame and screwed up my eyes as though expecting an explosion. If at that moment lightning had struck the forest and the earth had opened to release the serpent-crowned Furies from Tartary and all the demons of Hell, I believe that it would have appeared to me more natural than the silence which ensued.

As I watched this old, hairless, half-stripped woman, glowing with malevolent purpose, I realized for the first time in my life how one can become mad.

Quite dazed, I went and sat down beside the fire where I found Tchrachy, pale and silent, mechanically poking the embers. There was nothing for us to say. Each knew quite well what was in the other's mind.

An hour passed. Silence once more reigned in the monastery. Then Yong Rine returned, covered in mud, tired and empty-handed. Apparently the thief had mysteriously disappeared just as his pursuers were catching up with him. In Tibet this was quite enough to start a legend, and in the light of the fire I saw an expression of fear in the eyes of those around me. I felt that they could not look at me without thinking of the curse which I had brought upon them. These good lamas must have been fundamentally kind-hearted not to have flung out my ill-omened person then and there.

. . . . .

This incident had brought the tension of the monastery to a climax. That evening we barricaded ourselves in the house, the most outlying house of the settlement, waiting for a sudden irruption to kidnap or kill us. Our hosts, the lama and his mother, were prepared to share the night's misery with us, and were even tactful enough to keep up a show of gaiety. The man kept trying to teach me the names of the various objects used to prepare their meagre fare, and to please him I repeated the words over and over again, trying to copy the impossible pronunciation of the Ngolo tongue. His mother would watch for a smile to cross my face, and whenever she succeeded in cheering me she would chuckle delightedly and seize my hands in hers. The same performance had been repeated every evening since our arrival. She would retire very early to the room which she shared with her son, and whenever he stayed behind gossiping with my men, she would call him to order. He would reply with the irritability of a great boy of thirty, who thinks himself too old to be told off by his mother, but would only remain with us long enough to save his face before himself retiring to bed, leaving the fire to die down.

And so passed these aimless days, interrupted only by the tokens of goodwill with which I was constantly embarrassed. Not an hour passed but a lama would visit me, bringing me some present or other to express his sympathy. They would come singly or in pairs, shy and reluctant to cross the threshold, holding in their cupped hands paltry little objects which touched me so much that they brought tears to my eyes; little portions of butter, handfuls of salt, tobacco strands, dried fruits which I could neither swallow nor identify, and above all, what I most appreciated, large pots of curdled milk. They would remain there some time, squatting in front of me, replying to my gestures of thanks by noddings of the head, thoroughly enjoying having done a good deed, while I stood tongue-tied before them, unhappy not to be able to express my gratitude in intelligible terms. There was one fellow in particular whom I liked as much for his kindness as for his ugliness. He was a dirty little man and, unlike his fellows, wore a beard. He did not often shave, so his general aspect was that of an old discarded brush. He excused his frequent visits by bringing me only tiny presents at a time.

I had only one thing to offer them in return for their kindness, and that was the use of a pair of binoculars which Yong Rine was carrying at the moment of the attack, and which he had managed to save. When the weather was clear they would point the binoculars towards the grassy foothills of the mountains where the nomads' yaks were browsing, screaming with joy at seeing the distant animals brought suddenly close to them. Then they would make the inevitable experiment with the louse. Someone would search his clothes for one of these insects, of which Père Huc says that "in Tibet they are easier found than butterflies"; he would then walk a little way away, holding the louse in his hand, while the observer would try and focus the creature through the instrument. They were disappointed and could not understand why this device which could attract a yak had no power at all over a louse.

. . . . .

Finally, one evening, we left the monastery and the village of Dekho. Was it the sixth or the seventh day after our arrival? I must confess that I had not the smallest idea. I tried hard to count, but was no longer sure of the date.[1]

The incident of the stolen horse had brought things to a head. The following day we came to an agreement with the Grand Lama. The little bar of gold which I had with me was an additional asset. I had handed it over to a strange fellow, a Sino-Tibetan half-caste who, in spite of his Chinese name, was difficult to distinguish from the natives. The fact that he spoke Chinese was a source of joy to my men, who were delighted to speak and to be spoken to in a language which reminded them of Tatsienlou. He told me that he came every year to Dekho to buy musk-gland brought by the ibex-hunters to the monastery, which he then re-sold in Luho. The pleasure he showed in acting as negotiator between us and the lamas in the selling of this gold was a trait sufficiently Chinese to justify his claim to pass as such.

It was the gleam of the yellow metal which dispersed the Grand Lama's last doubts. He was becoming more and more anxious for us to leave, and had finally decided to appoint reliable men to escort us as far as the first Chinese gold-mine, near the valley of Serba.

The leader of the expedition had already paid us several visits, and I confess that no man had ever inspired me with more confidence. With his lined, bony face and beak nose, not at all of the mongol type, he resembled more a Calabrian bandit or a brigand from the Spanish Sierra than what he really was, a Ngolo adventurer. He was apparently the monastery's man of

[1] This explains why, on my return to China, I made a mistake in my report. I gave *September* 9 as the date of the attack, whereas it really took place on the 10th. I was able to establish this fact later.

action, the man whom the lamas summoned to perform acts of violence, alien both to their own priestly training and to the natural kindliness of their disposition, softened by monastic life. His business it was to deal with horse-thieves or with differences of opinion arising from violation of grass-lands. In a word, he was the monastery's warrior. With his felt top-boots, criss-crossed with blue, his wallet-shaped tinder attached to the girdle of his tchouba—a tchouba so long in the sleeve that the sleeve, which in the customary fashion he wore hanging down his side, almost swept the ground—the coil of his pig-tail wound round his fur bonnet and his long broadsword, inset with precious stones, he presented a remarkable appearance which would have enchanted Meissonier. One felt that he was ready for anything, and as careless of his own safety as of that of other people.

This was the man to whom we were going to entrust our lives. Although I knew nothing about him and although his brawling laugh was anything but reassuring, I put myself in his hands without a qualm. I was profoundly convinced that, by virtue of the law of hospitality, old as the world itself and valid in all latitudes, since the Grand Lama had appointed him as our escort we could have complete trust in him.

The organization of our flight was shrouded in such deep mystery that up to the last moment I had not the least idea what we were going to do; there were so many conferences behind closed doors, so many comings and goings, orders and counter-orders. Moreover, I was probably the only one kept in the dark for, in spite of the efforts they had made to keep it secret, our departure aroused such curiosity among the inhabitants of Dekho that it was impossible even to saddle a horse without the whole monastery knowing about it.

We left the hospitable home of the woman lama and spent our last night with the men who were to form our escort, lying higgledy-piggledy on the bare boards of the first floor of a building which might have been handsome, but which, like everything in Tibet, had never been properly finished and had grown old without ever having been young.

My companions sat up far into the night. They had killed a goat, bought in part-exchange for our gold, and were indulging in an orgy of cooking, slicing up the flesh to supplement our provisions for the road, an enthralling occupation which kept Tchrachy and Yong Rine from brooding on to-morrow's dangers.

Meanwhile, the Grand Lama had come to bid us farewell, and received us in audience in the largest room of the house. He came escorted by several monks, and made me a long speech, a veritable uninterrupted monologue to which Tchrachy listened with sustained attention. I thought he would never end, and began to suffer under the strain of keeping my attention fixed, although the most elementary politeness bound me to do so, for the priest never took his eyes off me. I understood vaguely that he was giving me irrefutable proofs of the innocence of the people of Dekho as far as the attack on our expedition was concerned, and, as he kept repeating the name of the Governor of Si-kang, Marshal Lieou Wen-hui, whom we were soon to meet, I also gathered that he was anxious to have us convey his loyalty towards him. The Lama had a shrewd head on his shoulders, and was taking advantage of the occasion to establish useful political relations.

Tchrachy replied on my behalf. I observed that his long speech in reply rang pleasantly in the Grand Lama's ears, and that he was exaggerating a trifle the intimacy of my relations with the Chinese pro-consul in order to gain the prelate's favour once and for all. Then followed other speeches from the high officials of the monastery. It was altogether a memorable evening, in which gastronomic concoctions alternated pleasantly with religious observances. Finally, the Grand Lama departed, leaving us to lie down on the floor just where we were, and endeavour to sleep.

I was convinced that we were going to leave at crack of dawn so as not to attract attention. Indeed there was a great hullabaloo in the early morning, which rather worried me; how could our departure be kept secret in this agitated monastery!

At least twenty times I thought the moment had come, but it was not till late evening, at the moment when I was least prepared, that they brought me my horse.

At the fringe of the forest I found my host and the little bearded lama waiting for me on the path to bid me farewell. It was a most informal farewell, thoroughly unoriental in its restraint. They gazed at me in silence for a while, and then, when I held out my hand to them, took it awkwardly in theirs, and with a sudden gesture of emotion, pressed it to their lips. I was so taken aback that I didn't resist, but gave them an affectionate squeeze on the cheek, gazing with a feeling of real tenderness at these men who were experiencing, probably for the first time in their lives, an irrevocable separation. And now that I was leaving this monastery in which I had spent the longest and weariest days of my life, I realized how deeply attached I was to the place and to the people.

. . . . .

My dear hosts of those sad days, the lamas of that lonely monastery! How touched I was by their simple, unaffected kindness and tact! I should probably never see them again, for by accident of birth they belonged to a world so far distant that they knew nothing of mine, and now that I had left them had themselves become as alien to me as though they inhabited another planet. Yet they had lavished upon me treasures of recollection. I did not know if their feelings of compassion were dictated by their religion, for their good actions were always accompanied by strange rituals. If it were so, then good luck to it. But I prefer to think that their kindness was innate.

I could only stammer their language, so their minds, incapable of deduction, disposed them to treat me as a child who had not yet learnt to speak. And this habit of theirs was so unexpected that it soothed me; a kind of distant echo of a mother's affection.

CHAPTER II

THE RETURN RIDE, September 1940

What a difference between these lusty tribesmen with their long moustaches and Virgil's languid shepherds, who spent their days playing on the flute or adorning their pretty straw hats with ribbons and spring flowers!

R. P. Huc.

WE had to make a fantastic ride to retrace our course over the district which divides the Tong from its tributary the Ser.

It would be quite impossible for me to mark out our route during those four days' journeys, sliding down into dark, wooded gorges, clambering up high passes, silhouetting ourselves along ridges, galloping across desert-like plateaux, already snow-covered. All I know is that the route was an absurd one and that our guide knew his mountain thoroughly. We changed valleys so often, we made so many unexpected twists that no one could possibly have followed our tracks in this maze, any more than I could have memorized our meanderings.

Our first day's journey only lasted a few hours. I imagine that the reason of our late start was to prevent some spy betraying our departure to any possible attackers. We merely followed the Tong valley, and I was surprised to find there several villages of well-built houses, and two important lamaseries surrounded by cultivated, terraced fields. So Dekho was the farthest upstream outpost of the settlers in the Tong Valley. I could therefore establish once and for all that Tibet is composed of two different worlds, the world of plateaux and herdsmen and the world of valleys and farmers, in one of which men live under flimsy canvas coverings, and in the other in heavy stone dungeons.

Ngolo Banquet

At nightfall we halted in a hamlet hidden away in a little valley-hollow. There, in a house where our hosts were apparently expecting us, the men indulged in a real orgy. Like all temperate peoples, the Tibetans are capable of swallowing at one go enormous quantities of food. Their usual fare is, however, frugal in the extreme. They are satisfied with two meals a day consisting of a few bowls of tsampa, with an occasional wheat pastry cooked under hot embers, and a little butter, the whole swallowed down with countless cups of buttered and salted tea. But when the occasion arises they gorge themselves with meat, finishing off to the last morsel all the quarters of whatever animal they may have killed, soaked in black broth.

The wild men who formed my escort—besides the leader of the expedition we had with us two lamas and two Ngolo laymen—when faced with this pile of victuals had become primitive beings, animals incapable of knowing when they had had enough. I myself, deprived of meat for several days and weakened by a diet which didn't suit me, felt an irresistible longing to get my teeth into flesh, and I differed from these Tibetans only in one point, that I was still able to stop at the right time. But like them I took great pieces of meat in my hands and tore them apart with my teeth; like them I wiped my greasy hands on my clothes with a gesture that had very soon become second-nature to me. It doesn't take long to strip off the varnish of civilization. When I look back I feel some astonishment at the recollection of that coarse banquet, but at the time it seemed to me quite ordinary, just as ordinary as the faces of the Ngolos round me, appearing and disappearing in the light of the flames.

I was already so accustomed to the world in which I was living that the memory of my own world was becoming more and more blurred. I forgot that I had once eaten off clean tablecloths, decorated with silver and flowers, with elaborate and refined manners which at that moment would have appeared to me absurd. But I had no time to dwell on such things. My complete moral and mental isolation, combined with a certain physical distress, left room for nothing but the instinct of self-preservation. It was this basic instinct, rather than hunger, which drove me to eat this repulsive food, and it was this instinct also which gave me strength to remain days on end in the saddle in spite of an attack of dysentery which weakened my resistance, with only one thought in my mind, to escape death.

I only regained contact with the past when I smoked. My pipes had been left in the saddle-bags of my dead horse, but Tchrachy often lent me his. It was a Chinese bamboo pipe with a copper bowl which held only a pinch of light-coloured, scented tobacco, the smoke of which was pleasanter to smell than to inhale and made me cough. But I always accepted the offer of this pipe, for the gesture of holding it and lighting it was the only luxury of my life.

In the moments when my mental stability returned I thought of hardly anything but Liotard. The image of France was growing gradually fainter in my mind. I was living among people who did not know that the world had gone up in flames, so it was difficult for me to drag back my thoughts to a conflict the immensity of which I alone knew.

. . . . .

I shall not attempt to describe the next day's journey. All I can remember is having spent a whole day in the saddle. An eleven hours' ride with only a midday halt of three-quarters of an hour to boil some tea and eat tsampa!

From dawn till dusk we pursued our uphill and downdale course, complicating our route, changing from one valley to another, riding upstream and downstream like a squadron of mad cavalrymen. The only people who saw us were the leaders of a caravan of about twenty-five to thirty yaks, who were preparing to camp. They stood there petrified. We were jog-trotting along and came up against them at a turning in the valley. They rushed to seize their arms, convinced that we were about to attack them. The men of my escort immediately shouted at them and stretched out their right hands with their thumbs jerked upwards in a gesture of friendship, a gesture which dates from time immemorial, probably long before the Roman games, and signifies goodwill, peace and pardon.[1] The travellers then lowered their rifles and stared at us open-mouthed, watching our extraordinary cavalcade, astounded that people who so greatly resembled bandits could pass by their property without wanting to grab it.

It must be admitted, that even in Tibet, where one is always meeting motley groups of travellers, wandering lamas, migratory tribes, pilgrims, traders, hunters, travelling bigwigs or raiding bandits, we presented a droll and sinister appearance. Our leader and his two cronies carried such an arsenal of guns, broad-swords and daggers that no one could have looked at them without shuddering. The lamas who should have provided the moral element of our party were themselves armed with large sabres, which contrasted harshly with their soft red outlines. Yong Rine had his musket with its magazine and the three rounds which remained to him after the fight. Tchrachy, bent double with the pain of his still unhealed wound, was the only unarmed man in this warrior group. As for me, I was inconspicuously armed with my revolver, but my bearded face and European clothes made me the most comic specimen of all in this motley band, which, if met unexpectedly, would have given the most seasoned warrior a nasty shock.

Those good caravaneers were the only people who had to suffer this encounter; thank goodness we avoided paths on which we might have met travellers, for it would only have ended in a fight. After this single meeting we saw nothing stir in this dead landscape but a herd of small antelopes, with curved horns, like chamois. None of us was tempted to take a shot at them, for the fear of the echo of the explosion was stronger than the passion of the chase.

[1] In Tibet, and probably in other districts of Asia, this gesture of holding out the fist with the thumb jerked upwards has always this significance. The same gesture reversed, that is to say with the thumb downwards, has, on the contrary, a malign sense.

Naturally we had left the cultivated lands soon after our departure. Even while still on the Tong's banks we had penetrated into a narrow, wooded gorge, quite uninhabited. There were fir-trees growing right down to the river banks, and fallen trunks lying across the water, wedged against the rocks and battered by the rapid stream; we might have been in a Canadian valley. For a whole hour our horses were no use to us, and often, in difficult passages, it was we who had to come to their rescue. Finally we were forced to take a little side valley and climb once more into the upper world, where there was a danger that the bandits were on the look-out for us.

In the last ten days the prairie had greatly changed. When we reached a high level I noticed that since the day of the attack the grass had had time to die. Tibet was gradually assuming that brown colour which it would preserve for months to come, awaiting next spring's explosion of life. There were no herdsmen's camps anywhere, and the marmots had already gone to ground and no longer disturbed the air with their strident cries, that air so thin that even birds cannot exist in it. When we were able to get a distant view we saw that there were many more snow-covered peaks. Autumn had come at last, the season which we had awaited so impatiently. It was miserable to think that neither Liotard nor I would be profiting by the lovely, clear October days ahead, nor continuing our journey northwards over the vast, unknown spaces of this Ngolo country, whose veil we had barely lifted.

Nevertheless, next day there was a final hailstorm, which, by a coincidence which strongly impressed all the eight members of our party, descended on us with extraordinary suddenness at the very moment when we were ascending what seemed to me the highest pass of our excursion and the crossing most feared by my Tibetans. I had lost my altimeter but from the penetrating cold on this ridge and from the difficulty I found in breathing I calculated it to be 17,000 feet high.

We had to tackle the ascent and my companions began staring about them in that way of theirs which I knew only too well. I felt in my heart the double pain of the height and of anxiety, and I could not help making sad comparisons between this pass and "the other", forgetting that all dead places of the earth are more or less similar, where the bony structure of the globe has torn away the outer layer of life, baring everywhere the same mineral carcass, enduring evidence of ancient geological ages.

Here the strata were vertical towards the sky, broken, and forming rocky crags; these crags so hindered the progress of our horses that we had to dismount. At the very moment when the Tibetans, with their rifles cocked and dragging their horses behind them, started praying out loud, the thunder suddenly burst and enormous hailstones spattered the ground, stinging our hands, causing our horses to shy, and covering the whole landscape with ice.

At the head of the cavalcade the leader made a quick, steady ascent, his eyes fixed on the jagged line whose outline was blurred against the heavy sky. Climbing with his rifle held vertically in both hands, he looked like a tight-rope dancer leading a charge. The others followed, their faces tense, trying to see through the rock face to discover whether it concealed enemies or nothing. At last the leader reached the ridge and his figure was outlined against the storm clouds. Dropping his rifle he turned towards us with a gesture of reassurance, and my companions gave a shout of joy. I believe that I, too, for the first time, actually cried out with relief. Perhaps, at that moment, I was feeling the strongest emotion of my life, so true is it that the expectation of danger is worse than the danger itself. I felt that in crossing this pass to the tune of this thunder and this angry sky I had finally exorcised my ill-fate, and that the wrathful gods would henceforth leave us in peace.

At the other side of the pass the plateau extended its ocean-like swell, at present covered with a carpet of snow, and although the hail had soon thawed the sky appeared black by contrast. We should be obliged to travel across this dazzling carpet, along these foothills which would offer poor cover, and our tracks would remain relentlessly clear like roads marked on a map. So we increased our speed. Our horses were beginning to feel the cruelty of this perpetual scenic-railway journey, but we nevertheless spurred them on. My horse was poor, and I sometimes found myself several hundred yards behind the others. But my escort never forgot that the purpose of the journey was to deliver me into safe hands, and always waited for me. And yet I could see that they were nervous at feeling themselves thus exposed to view for several miles round. It was strange to think that these men who did not know me and who in other circumstances would not have hesitated to fall upon my caravan would, in the event of attack, have fought bravely for me, stranger though I was. The ethics of nomadic peoples only take account of particular situations, and they have no intrinsic conception of right and wrong.

For another ten hours we pursued our third day's monotonous and weary course. First came snowfields, then valleys, then more valleys and then more snowfields, and at last, deeper down, we reached a district of desolate, stony plateaux and a black, stagnant lake. Sometimes there was no life at all around us, and sometimes we found the prairie yellowing and without flowers; in the small valleys we sometimes rode through brushwood and when we descended very low the forest would appear in patches. Everywhere in this callous landscape we felt ourselves intruders, for this territory had escaped the bondage of man. If by hap creatures of our kind had passed this way they had left no trace.

Only once did we see two of our fellow-creatures, horsemen faintly visible on a distant ridge, and from the anxious expression in the eyes of the Tibetans I realized that man is the greatest enemy of man.

Night had almost fallen and the fog was already beginning to blur our outlines, but we were still riding along in silence, like shadows. I had begun to feel tired at the seventh or eighth hour of our journey, but after that bad moment I felt a renewal of vigour and began to believe that I could remain indefinitely in the saddle until my horse collapsed. My bruised loins gave me no pain; only my hands blue with cold were still aching.

Finally, as on the previous night, we camped somewhere in a hollow, under some trees. The Tibetans had lit a fire to boil the tea, and immediately afterwards trampled it out lest its light should betray our presence. No one spoke above a whisper. If our presence had been noticed in the daylight we must take care not to betray ourselves through any rashness. We had not camped at the bottom of the valley and our horses were picketed to trees close beside us, so no one could discover us in this thick darkness.

The chief showed a concern for my comfort which contrasted strangely with his bandit appearance. He cut up a bundle of twigs to make a bed for me and, in spite of my protests, insisted on covering me with his skin coat which smelt of animal and human fat. Then, laughing and thoroughly pleased with himself, he went and snuggled up beside his companions.

It would be difficult to say what motives inspired him to behave like this. I knew that he had received orders to protect me, even at the risk of his life, from a man whom he regarded socially as his chief, but the tenderness with which he carried out his duties was entirely spontaneous. It was an example of the contradictory nature of this extraordinary people, rough and yet gentle, one of the most attractive people in the world whom, in spite of the tragedy that I had suffered among them and at their hands, I shall never bring myself to hate.

Above me were the starry heavens and the bole of the tree which sheltered us and which rose tapering into the sky like a ship's mast. It made me think of other nights spent in the open, stretched out at full length on the deck in the tropics, gazing up at the firmament which glided from port to starboard, from starboard to port, round the motionless cross-tree at the mast-head, long, long ago when I was a young sailor, in other words when I was different to what I am now.

. . . . .

On the fourth day of this exhausting journey we reached the Chinese gold-mining camp. That very morning we had seen on our journey several herdsmen's tents, with horses grazing at large around them. The settlers had looked at us with surprise, but had not rushed to take up arms, which seemed to me astonishing. We were entering one of those havens where, protected by the little Chinese garrison, the inhabitants have developed a sense of security. The mission to the Ngolos was drawing to a close.

Towards midday we reached a small ravine and I saw ahead of us the Chinese camping-ground. And now the men who had protected and cherished me for four days, who, if need be, would have fought and even died for me, left me without a word. Reining in their horses, they signed to us to dismount from ours, which they had to take back to Dekho with them; then, turning in their tracks, they galloped away, hardly bothering to give us a farewell wave. Their friendship seemed to have ended with the contract which bound us. We had become strangers to them once more, and, who knows, perhaps even enemies. Moreover, they did not fancy making contact with the representatives of an authority who might have asked them to account for, or at any rate describe an incident which they did not wish to be mixed up in. I watched the departure of these men who had kept so faithfully to their word, and as their figures grew faint I thought that nothing really distinguished them from any band of robbers. Thus ended the story of my life with the Ngolos, this people who so strangely combined brutality with proofs of kindliness and good nature.

. . . . .

I entered the main hut of the gold-washers with the joy of a shipwrecked man stepping on to the bridge of the boat which has fished him out of the sea. I once more had the feeling of being in security; I regained the sense of lawfulness as opposed to the law of the jungle, of despotism and fantasy run riot; I forgot the primitive mentality of the tribesmen of the plateaux who still live in a past which dates from the era of the great barbarian invasions. Once upon a time I had smiled at that label "barbarian", which the Chinese apply loosely to all aboriginal populations in the confines of their Empire, but now I understand it only too well. As a civilized man, accustomed from long ages to life in organized society, I suddenly discovered that, in spite of ethnical differences and the barriers of language, these humble Chinese, petty officials, workmen and soldiers, were men who also had a long civilization behind them, that they were cells in a great organized body to which I, lost cell that I was, felt attracted as though by some biological law.

This camp, which consisted of twelve badly-constructed huts, was nevertheless very poor. It was inhabited by thirty or forty settlers, living higgledy-piggledy and in rather wretched conditions, engaged in washing the river sands. I perceived, with pity and amazement, two little civil servants dressed in skimpy European overcoats, who looked as absurd in this country as two Tibetans in tchoubas in the Champs-Elysées. The bric-à-brac of objects with which they were provided—blankets, kitchen utensils, china teapots, leather boots, buttoned clothes and enamel basins—appeared to my admiring eyes like a priceless treasure. I gazed with particular excitement—may the reader excuse my childishness—at a plain toothbrush in a glass, and could not decide which caused me greater pleasure, the brush or the glass. These Chinese mandarins, with their customary politeness, invited me to sit down at the table, for there actually was a table, and offered to share their meal with me. I felt a real happiness overwhelm me, a happiness which reached its climax when the kindly men offered me my first cigarette. Our tobacco fumes mingled with the scent of opium, indicating that these exiles derived comfort in their boredom from the compassionate juice of the poppy.

That same evening, after a three hours' journey, we reached the valley of the river Serba, and the hamlet of Lien Po Chan. I took up my abode in the house of the captain who commanded the district of Serba. For the next few days, in this stone dungeon, I was to feel total security, to recover the sense of my own personality, and to realize my grief.

. . . . .

M. Tcheng, the commandant of Serba, was a very worthy man, who appeared utterly bored in this lonely outpost. Like Sahara officers on the fringes of the desert he had given up shaving, and the few hairs which hung down at the corners of his lip and on his chin gave his face the appearance of a Chinese theatrical mask. With the help of a few pile carpets he had managed to give a cosy atmosphere to his little office on the second floor. It would be easy to make those solidly-built Tibetan houses comfortable, with their massive walls and thick floors, but the Chinese definitely do not trouble much about improving their living conditions. M. Tcheng's assistants, civil servants, officers and accountants—there was even a wireless operator—lived in a dormitory in a large room on the first floor, and I believe they enjoyed this higgledy-piggledy mode of life because it suited their gregarious instincts.

I, too, driven by the desire to mingle with my fellow-men, left the house in which we had first taken up our quarters and where I was alone with my two men and went and slept in the dormitory. I was disturbed, sometimes far into the night, by the noise of Mah-jong counters clacking against the wooden tables, but at least I was not alone, and all the Chinese round me were full of concern for my welfare; a concern less spontaneous and more formal than that of my old friends the lamas.

Moreover, now that I was once again among people who seemed to have other things to do besides consult the fates, bow down and tremble before deities, I felt the spell of Tibet loosen and freed myself from its baneful influence; I began once more to react otherwise than by instinctive or emotional impulse. Chinese realism was gradually lifting the nightmarish cloud under which I had for so long existed.

The activity of my new companions was, it must be owned, not excessive. The captain was content to give a few orders here and there and to listen to his subordinates' reports. The latter would draw a few characters on the account books or weigh pinches of gold dust on minute scales. Some busied themselves with distributing foodstuffs to the cooks, others with bringing in the last barley harvest, whose sheaves, hanging from the large lofts which topped the houses, gave them the appearance of immense ricks resting on stone sockets. The rest of the time—and there was indeed a lot of time to spare—was spent idling in the courtyard when there was not too much mud, cracking jokes or gossiping with the few Tibetan peasants who visited this Chinese village. But inexacting though they were, the activities of all these men revealed their function in regard to the community and thus bound them to the world where work is the *raison d'être* of man.

Although Lien Po-chan is a military outpost, facing unsubdued territories, rather like a garrison in South Morocco, there is no display of militarism. The conquerors have built no fortifications, not the smallest redoubt to resist an attack; the soldiers neither drill nor form guard, and their arms remain in the rack. No one seems to fear anything. Everywhere in the marches of the Empire the Chinese domination has that mild, lazy appearance which so greatly surprises Westerners, whose habit is to establish their rule on the basis of impressive displays of metal and steel. Nowhere does the deep-seated pacifism of the Chinese appear so clearly as in these colonized districts. It is much less a question of subduing rebellious populations than of trading with them, and of benefiting by the resources of the country. In China, therefore, the soldier is only a walking-on part who cannot be dispensed with but who is never put in the limelight. That is why there are more robes than dolmans to be seen in Lien Po-chan.

This Chinese good-nature does not prevent hard knocks. Every now and then an entire outpost is massacred. No one appears over-concerned. Others, as carefree as their predecessors, are sent to replace the slain, and they pursue their peaceful trade with an equal disregard of danger. This form of peaceful penetration has a certain grandeur about it. That in this valley of Serba a handful of little officials, half-trained soldiers and coolies should be able to exist surrounded by these armed and rather warlike tribes, shoulder to shoulder with powerful lamaseries which could easily rouse the population against them, gives a good idea of the courage of the Chinese and of their colonizing power, both virtually unconscious qualities, for not one of these men, when he returns to his province, would think of posing as a hero. Like true orientals the Chinese believe rather in the solidarity of ties formed by community of interests than in the prestige of heroism. So their chief aim is to gain the support of people likely to profit and to let them profit by economic advantages. Unfortunately, one sometimes finds an unscrupulous official who takes advantage of his responsible position to get rich quick; the exasperated natives then resort to violence, and we have bloodshed.

The day after my arrival I was prompted, slightly by respect but chiefly by curiosity, to pay a visit to the queen-ruler of Tsatsa.[1] She lived on the left bank, a little way upstream from the monastery of Yu Gomba, whose buildings rise in tiers up the slopes facing the hamlet. She seemed to be on the best of terms with the Chinese who accompanied me, and they for their part showed her the respect due to a person of quality. Her well-built residence was situated at a bend in the river, and she received us very politely in the large living-room on the first floor, offering us a collation consisting of buttered tea, fried cakes and ham. I might have been sitting in any prosperous French farmhouse. This little country queen was a rich landowner of handsome appearance and still young, adorned in turquoise and corals. She seemed to be quite resigned to the overlordship of these mandarins who had come to visit her, and did the honours of her country seat in a graceful and unaffected manner.

Once more I admired the free and easy attitude of Tibetan

[1] The original name for this place. Lien Po-Chan is the name which the Chinese have given it.

women, so different from the stilted, timid behaviour of their Chinese sisters, accustomed by centuries of constraint to efface themselves and remain silent in the presence of the male. Tibet is a feminist country and polyandry is still practised in certain districts, although it is difficult to decide whether this custom, which is gradually fading out, is a survival of matriarchy, traces of which are to be observed in Tibetan society, or whether it is the result of the economic conditions which prevail. The area of cultivation is not very large and cannot well be increased, so the fact of several brothers marrying the same woman and thus legalizing a household of three or even four prevents the partitioning of property.[1] The same explanation might apply to the herdsmen, who would in this way avoid a division of the herds, although in their case the mere promiscuity of tent life is already conducive to sexual intercourse. Whatever the truth is, the mistress of a Tibetan household, being more or less the centre of the family, the one who begets children for the community, acquires thereby an independence of attitude and behaviour in regard to the community itself much more similar to that of European women, who, it must be admitted, have themselves a tendency to polyandry, than to that of her Asiatic sisters, who tend more towards polygamy.

. . . . .

I spent the time before the midday meal enjoying the sunshine, because, in this stronghold with its thick walls slotted with windows which are really loopholes, the temperature hardly changes; at night it is higher than the temperature outside, but lower in the daytime.

My curiosity had been aroused by a strange little scene which had apparently attracted a dozen or so idlers on the look-out for diversion, and I joined them without exactly knowing what was going on.

The protagonist of the drama was a wretched Chinaman,

[1] Prince Peter of Greece, Tibetan explorer and specialist in polyandry, favours this economic explanation.

probably a coolie from the mines, and he did not seem to be pleased at being thus put into the limelight. Ragged and miserable, he was trembling like a condemned man on the road to the scaffold. I looked on in amazement, wondering whether I was to witness an execution, for one of the officials in charge was armed with a sword. But when I saw that it was not the usual large executioner's broadsword which still symbolizes justice in the remoter districts of China, but merely a crude meat-chopper I felt slightly reassured. They could not be going to sever the poor fellow's neck with a kitchen-knife! Besides, the presiding officials treated him with a kindly condescension which alone would have comforted me, apart from their jollity and the amusement of the crowd. So what the devil were they going to do with this man, seeing that all their jokes and encouragement failed to relieve him of his terror?

After long hesitations and withdrawals the snivelling wretch eventually held out the forefinger of his right hand and they tied a piece of string round the top joint. Then, gently but firmly, they coaxed him to place this finger on a fencing-post while another man drew the string tight. It was not such an easy matter. The poor man kept drawing his hand away instinctively, which provoked great outbursts of merriment among the crowd and more and more coaxings from the officiants, who seemed to think he was making a great deal of fuss about nothing. Finally, more from exhaustion than from conviction, he relaxed for the fraction of a second. The chopper rose, descended upon the finger and off jumped the joint. Everyone looked relieved and happy, even the victim himself, who, after all, was thankful to get the painful operation over. The executioner smiled and apparently began telling his victim that he had been as good as his word, and that it was only a short trial after all. For two pins he would have asked him to thank him.

Tchrachy, who had been watching with the greatest interest this everyday rough-and-ready execution, explained to me that the man was a hardened thief who had been sentenced to suffer a mutilation. I was a little sceptical about this, for I could not see any explanation for this summary amputation.

To tell the truth, the whole thing had taken place in such an atmosphere of good-humour that it didn't occur to me to feel angry. The punishment of a poor homeless devil who has an over-inclination to pinch things is not an inexpiable crime if it only consists in the loss of a finger-joint.

Moreover, it is difficult to see how, in these remote districts, justice can be administered otherwise than by corporal punishment or by the *lex talionis*. The penalty of imprisonment only hurts the magistrates who inflict it. The delinquents themselves, having no great needs, do not suffer much from the restriction of their freedom, which is pleasantly counteracted by the pittance they receive in exchange for the various little tasks allotted to them.[1] On the other hand the local mandarins, who are usually poor, are obliged to support them out of the administrative funds, so they are not keen to have prisoners on their hands for months at a time. This is probably why they favour corporal punishment which costs nobody anything.

And yet I could not exactly see what there was comic in this scene. In other circumstances the agonized expression on the man's face might have made me laugh, but not at the moment when he was about to suffer. I know that the sense of the ridiculous varies among different races, but to burst into guffaws at the sight of a man in pain was really going a bit far. I realized that after this exhibition I should never be quite at ease with my kindly hosts.

. . . . .

I was soon to discover that in China laughter is often used to hide emotion. Every time during the next few days that we

[1] A regard for truth prompted me to describe this picturesque scene. But one must not generalize from this story and conclude that corporal punishment is practised everywhere in China. The modern Chinese code has abolished it entirely with the exception of the death penalty. But China is such an immense country that we must not be surprised to find survivals of the past in the more outlying provinces. Moreover, the incident related happened more than five years ago. Besides, think of what has happened in Europe since then!

told the story of our journey to mandarins or other Chinese people we were received with what sounded to my European ears like exclamations of gaiety. These good people would listen smilingly to our dismal tale and at the most tragic passages inevitably burst into shouts of laughter. Slightly embarrassed to say the least, I at first took this manifestation for a nervous habit peculiar to the individual I was addressing. But soon the evidence told against it; they all reacted in the same way, so it was a characteristic common to them all. But I had nothing but praise for their treatment of me and for the real delicacy of their attentions, so that, in spite of their apparent indifference to the sufferings of others, I never imagined that they enjoyed hearing about my misfortunes.

I very soon realized that Chinese manners forbid emotional outbursts of any kind. Anger, the coarsest of all emotions, must never be displayed, and it is the same with pity or grief. This pretence derives from a sort of shyness which compels these people not to inflict on others the spectacle of their grief. No Chinaman has ever announced the death of a friend without a smile on his face.

A long time after, when I was settled in Chungking, I called on a Chinese minister for whom I have the greatest admiration and affection. One of his sons had been killed two months previously in air-warfare against the Japanese. He received me in apparently the highest spirits, and while we talked about his loss the smile never left his face. He only began to look serious when we started talking about the weather. On another occasion I was interviewing a young Chinaman whom I wished to engage as a secretary. I happened to ask him if he was married at which he burst into peals of laughter and said: "No, no, my wife's dead."

Our European good-breeding insists on an entirely different code of behaviour. Whenever we hear of the death of someone who doesn't concern us or even of a stranger we put on what we call a "becoming expression". In both cases, therefore, it is pretence. I must say, in favour of the Chinese system, that there is less hypocrisy in pretending to be gay when you are sad than in pretending to be sad when you are quite indifferent. But I, for my part, shall not suffer any less for having to work up a pleasant smile in order to appear ordinary; when, in proportion as I regain contact with the world and the danger lessens, the thought of Liotard's death becomes a painful, throbbing wound.

. . . . .

In spite of the sympathy shown me by the wireless operator, a charming young fellow who knew a few words of English, I left Lien Po-chan still in complete ignorance of the happenings of the great world. He showed me with pride the wireless station, but it did not work and no one seemed to worry very much about it. If these people were at all concerned about their safety, they obviously relied more on themselves than on the help of wavelengths.

It was now the end of September and since July I had had no news at all of France or of the war. I was on the point of regaining contact with actualities—for I was beginning to think that I should come out of this adventure alive—and the prospect of being suddenly presented in a lump with three months of history added a fresh agony to my private sorrow. It was as though I were living in another age, and I was probably the only Frenchman who knew nothing of the fate of my country and of my compatriots. But I was now to return to the plains where men live, to renew contact with the present, to grow several centuries younger, and although I desired this the idea frightened me a little. I did not know where lay the worst nightmare; in the solitudes behind me which I was about to leave, or in the countries before me, ravaged by war.

Once more I was bidding farewell to fellow-men who, in spite of differences of colour and race, had given me the warmth of human sympathy. I think it was *September* 24. M. Tcheng had forbidden me to travel alone with my two men without an escort, so I had to wait a few days for a Chinese caravan to be formed, consisting of minor officials and gold-washers, which was proceeding to Luho under the protection of two Tibetan oulas.

I was about to retrace my steps to Tatsienlou, no, even farther, as far as Chengtu, following the very same route upon which Liotard and I had so gladheartedly embarked a few months previously. I was to see our previous camping-sites, places where we had been happy and where we had been uncomfortable. I was to spend the night sometimes in the very house, the very room which I had shared with my companion; I was going to meet people whom we had met together. It was to be for me a long and painful calvary.

The pain started during the first minutes of the first day, at the last house of the hamlet. Tchrachy had gone in and come out again, looking thoroughly upset; he was dragging our bitch Roupie by the scruff of her neck, Roupie whom we had entirely lost touch with since the attack. Poor little Roupie, who had vanished and who now appeared to us like a ghost on the threshold of this house! She had grown so thin that she was hardly recognizable, and her dog's eyes were glazed and lifeless. I don't think she recognized us; our voices only made her tremble and she ran away with her tail between her legs into the shadow of the stable. Did her animal mind bear any recollection of the horrible scene which had interrupted her carefree journey and her mad chasings after marmots? She must have been the only witness of our friends' last moments. What had she done since then? Her instincts had probably prompted her to keep with the herds rather than with the men, and when she saw the men either dead or scattered, she had no doubt stayed beside the horses and the yaks up till the moment when the bandits had chased her off. She had probably then sniffed her way to this village with its stone houses, and we decided to leave her where she had found shelter. Poor Roupie! Her fate had been for a few days linked with ours, and she had never learned what we had wanted of her.

Callot would have liked our new caravan. It was a motley assortment of people who had assembled with their old clothes and their strange equipment to travel in company across the lonely, hostile district which separated us from Luho. It was a regular ambulant museum of old-fashioned weapons. It contained everything, from clumsy Mauser revolvers slung over the shoulder and decorated Chinese-wise with tassels, to old flint muskets, up-to-date carbines and large broadswords. The thirty-odd people who composed this body were carrying with them all their worldly goods, so their horses and mules were weighed down under absurdly heavy loads. The two Tibetan guides, armed with forked guns and mounted on handsome steeds, looked with warriors' disdain upon this picturesque group of overseers and workmen, trembling for the wretched pittances which they had amassed in the mining district. And yet the danger was not to be despised; up till reaching Tatsienlou and even beyond until they reached the motor-roads, these wretches risked losing in one second the fruits of months or even years of toil, lucky to be found lying at the edge of the pack-trail, despoiled but still alive.

Our journey was, however, uneventful. I shall not retail it. What point is there in recalling in detail those long, dull stages, the only purpose of which was to progress. On the third day I reached Luho, where for two days the mandarin entertained me in great state. Then I started off again. I spent one night in the Charatong mission, where Père Ly tried talking to me in Latin, the only language in which, at a pinch, we could communicate our thoughts. Very early next morning I mounted Père Ly's own horse and, after a ride of almost forty miles, arrived at nightfall at the mission of Tao-Fou where, in a state of wild excitement, I fell into the arms of Père Yang. At last I was able to talk French again, a joy which for the last month had been denied me. Moreover, I was to form a deep friendship with this young Chinese priest which the passing of the years has not lessened.

. . . . .

What followed was only of interest to myself. It is of no importance that I stayed in this monastery or that, that I took one trail or another, that bandits again occasioned us some alarm, that it was very cold crossing the passes. I was becoming once more a man like other men, though when I saw myself in the glass I noticed that I still looked a bit wild. I was escaping by degrees from the grip of the adventure, and I felt my own self returning, that self which I had to some extent forgotten, and which was to mingle once more with the crowds whose call I vaguely felt.

Finally, on a radiant October morning I reached Tatsienlou. I crossed the bridge where, four months previously, we had said good-bye to Père Doublet. Since then we had not met a single man of our own nationality, and for a whole forty days, since Liotard's death, I had lived alone among Tibetans.

I was moved to see once more the familiar and dear faces of the priests of the French mission to Tibet. Mgr. Valentin expressed his grief at my friend's death, and at the same time told me that the mission itself had suffered a cruel loss. On *September* 17, in the village of Pamé through which Liotard and I had passed in 1937, a French missionary, Père Nussbaum, had been assassinated by brigands. Thus, in less than ten days and at several hundred miles distance, two fresh names had been added to the roll of French martyrs in Tibet. One of these men had died for his faith, the other for science, both for a disinterested cause and for the greatness of their country. For the priest's sacrifice and that of the scientist were united in France, the country which had given birth to them, whose soil had nourished them and whose soul had formed their character.

The news I heard from France in that October of 1940, would have driven me to despair, had it not been for the presence in London of that mysterious Taï Ko Lo. I learnt at last that Taï Ko Lo was General de Gaulle. It was he who was to be my guiding light in the topsy-turvy world to which I had returned. I found to my great surprise, on escaping from this cruel adventure which had so dazed and bewildered me, a clear path ahead. All Frenchmen owe this man a deep debt of gratitude, but I more than anyone; I owe him the fact of not having sloughed into despair, for in addition to my personal misfortunes, I had to bear such a burden of horrible news learned accidentally from conversations with the missionaries that, without the hope which I placed in him who personified France, the ordeal would have been too great for me.

As I set off on my descent towards the great plains of Western Asia, I knew what was awaiting me. I knew that I was returning to the great world torn by passions not mystical but human, that I was to see the cities of Sseu-tchouen bombarded by the Japanese, and the fever of war. I knew too that I was going to meet Doctor Bechamp once more, and that this man of genius would be not merely a great consolation to me but also an incentive to carry on the struggle, for he himself was preparing to rally the Free French.

What I did not know was that I was destined to remain so long in China, among these Chinese people whom I was to grow to love, and that I was to help in the astonishing strides of this great country. But if I had known that this country was to harbour and become a second home to me I would have been less afraid of the exile which lay ahead, and which I felt would be a long one.

During all the last days which I spent in Tatsienlou there were constant earth tremors. It was as though the soil of Tibet intended, up till the very last moment, to fulminate against me for having tried to violate its mystery.

However, before leaving, I witnessed an even more serious sacrilege, the arrival of the first motor-car in Tatsienlou. The Chinese who, in the course of these years of war, have proved themselves great road-constructors, had, by dint of enormous efforts, finished the road which for the first time led to the roof of the world. It is true that this car was a small one, and that they had had to carry it several miles, but the hooting of its klaxon sent the crowds of lamas and Tibetans, who had assembled to watch this important occasion, scattering in all directions. But I know that, since then, many vehicles have traversed the same route, and that farther northwards another road starting from the Kou-Kou-Nor has penetrated bit by bit into Tibet and has reached the district of Jyekundo, near which Dutreuil de Rhins was killed. This unhappy pioneer, this French explorer, had planted the flag of progress in this spot where he had died, and fifty years later the Chinese advance-guard sighted it. Such sacrifices are not in vain.

It will probably take less time to reach Louis Liotard's tomb, in its grim solitude, and to mark the exact spot where he died. Mediaeval Tibet will not long resist the relentless law of evolution. I have loved this country for its strangeness and archaism and shall no doubt be weak enough to deplore its passing. I shall be one of the last Frenchmen to grieve for it, for, since my last expedition, Père Doublet is dead, and Mgr. Giraudeau, the doyen of the Tibetan missionaries, is also dead. Père Burdin, who made his first journey in our company in 1937, has also perished on the forbidden territory of Tibet.

It is dangerous to wait too long before writing one's recollections, because in the meantime many people die.

But, whether they died by sword or bullet, of illness or old age, were they priests or scientists, peasants or aristocrats, bore they the name Dutreuil de Rhins, Nussbaum, Burdin or Liotard, all these Frenchmen have given their lives for an ideal, and the apparent uselessness of their endeavour makes the very greatness of these men and of the nation who bore them.

EPILOGUE

A FEW months after my return to China I learnt from Père Yang that it might be possible to rescue an important part of our documents and of our material. At the same time he told me one or two other interesting facts about the attack on our expedition. Apparently the thieves who had stolen our yaks and the men who had killed Liotard and Tze formed part of one single band, and they were probably the same as those who had come sneaking around under our men's tent, and with whom we had nearly had a scuffle (see Chapter 6: Chortaintong). The object of the yak-stealing was merely to find out the extent of the magical powers which they attributed to us. So I was confirmed in my view that the attack was no mere act of plunder, but that it was also a question of the supernatural. It is possible that our work of anthropological measurements had confirmed the idea that we were practising sorcery. Père Yang added: "By a miracle you escaped with your two Tibetans, and your attackers were terrified. They picked up all your equipment which is still intact, because none of them dared touch it."

So my escape unharmed with my two Tibetans from an ambush so carefully arranged that we ought by rights to have been killed had strengthened the brigands' conviction that I was endowed with magic powers. It is possible that they also ascribed to this magic power the apparent calmness with which I moved from place to place during the affray, whereas it was entirely due to the effect of altitude.

In any case I wrote at once to Père Yang to ask him to take all steps to recover, if not our equipment, at least our documents. Doctor Béchamp, the French Consul in Chengtu, in his turn, asked the Chinese authorities to be good enough to instruct the Si-kang officials to help in this recovery. Both parties acted with all possible speed. All the tribes were warned, from Chinese mandarins to Tibetan chiefs, and a small expedition of about fifty riders was organized. Seeing that the odds were against them, the chiefs of the clan to which our bandits belonged decided to disclose the spot where our things were, and even handed over a few men who were suspected of having taken part in the attack. I do not know what happened to these men.

Unfortunately, the material passed through so many hands before reaching Père Yang that almost the whole lot disappeared. None of our valuable apparatus or instruments were saved. So as far as that was concerned the loss was total. On the other hand the various observations we had noted down had been preserved owing to the fear which they inspired, and to their apparent worthlessness. I recovered the greater part of them. As was the case in 1895, after the attack on the Dutreuil de Rhins mission, when the documents had been recovered some months later, many of our films were fogged because the thieves had opened the rolls. I only rescued the photographs which had been already developed at the time of the attack and two 5-millimetre movie reels. The 12-millimetre reels were all spoilt, and also our colour films.

I wish to express my gratitude to Père Yang, Governor Lieou Wen-hui, and the Chinese officials, thanks to whom Liotard's death and our efforts in this tragic expedition have not been in vain.

. . . . .

The findings of the expedition, as set down in our notes which had been so unexpectedly preserved, were not without interest. They made it possible for me to enlarge the map of the Ngolo country, thanks to Liotard's survey of our journey of nearly a hundred and twenty-five miles in unknown territory. Later on I was able to use this survey to make a general sketch-map extending from the eastern boundary of the Bayen Khara mountains to the Mynia Gonka massif. This map has been published in the U.S.A. in the *Geographical Review*. So the expedition has thrown some light on the orography and hydrography of the district, as well as on its general morphology.

In addition we took a great number of photographs in the course of our three hundred and sixty-odd miles journey from Tatsienlou to the Liotard Pass, as well as several panoramic views and a few sketches.

Unfortunately, our ethnographic notes have disappeared, but our anthropological observations have been saved. An essay derived from observations made upon fifty or so Tibetans of the Hor communities is in the process of completion and will be published shortly.

. . . . .

Liotard's body still lies on the pass which bears his name. Yong Rine told me that with the help of the lamas he had made a grave for him by piling stones over his body. I begged Père Yang to ask the Tibetans to go and engrave his name and the date of his death on one of the rocks of the pass. Père Yang had intended to remove his remains to Tao Fou and bury them in the Catholic cemetery of the mission. But it has not yet been possible to carry out this plan. Personally, I would rather he remained on the spot where he met his death, a milestone in the discovery of the earth.

GEOGRAPHICAL NOTES

The country of the Ngolo-Setas in the district of the upper basin of the Tong.

(THE country of the Ngolo-Setas which is the subject of this essay is the country which we have in part explored, whose territory coincides with the drainage area of the Tong in its upper basin. The habitation of the Ngolos is probably not limited to this district. The territory of this important human group—I hesitate to use the word tribe, which would convey a too-exact sense—probably extends farther to the north, as far as the districts watered by the great bend of the Houang-Ho. I adopted the name "Country of the Ngolo-Setas" to denote the upper basin of the Tong, in order to avoid any possible confusion. The Setas form an important tribe whose country does not extend beyond this basin and who seem to consider themselves as belonging to the Ngolo group.)

The country of the Ngolo-Setas lies between two important systems, the extreme of the eastern end of the Bayen-Khara mountains on the north-west, and the complex formed by the chain which extends from the Tao Fou mountains to Mynia Gonka, and by the Sseu-tchouan Alps on the east and south-east. The nearer one approaches to Sseu-tchouan, the more the main trend lines of these mountain systems tend to deviate to the south. The Bayen-Kharas seem to have a well-defined west-east orientation. The Tao Fou-Jara chain sets from the north-west to the south-east. The Sseu-tchouan Alps, prolonged southwards by the Mynia Gonka massif, although little known, appear to have a general orientation north-south.

Between these systems, many of whose summits exceed 19,000 feet, lies the Ngolo-Seta country, a plateau or rather a peneplain of comparatively low altitude, with undetermined geographical boundaries, whose highest level is, at a generous estimate, 16,000 feet.

The word plateau should not be taken literally. These highlands are far from presenting an entirely tabular appearance. In many places, far from the big rivers, their relief has the classic aspect of a very ancient abraded surface, but everywhere else the district is in full process of rejuvenation, of which the Yalong and the Tong are the active agents. The great depth of the valleys of both these rivers below the plateau surface gives a strong quickening force to their tributaries, and this facilitates the process of headward erosion.

Thus the plateau is attacked from all sides. But where it still exists it is a landscape of smooth undulations, with pure, unbroken outlines, without any appreciable break of slope. It is indeed the "green awning" which the French explorer Jacques Bacot speaks of. It is barely broken here and there by the outcrops of the rock *in situ*, perhaps edges of strata tilted up at 45° or perhaps tips of pinnacles, probably granitic. The headwaters, on the plateau, at an altitude of 14,700 feet, are merely meandering brooks winding over a sodden soil, regular peat-bogs which the wet summer maintains until after the thaw, when these lands would be able to drain. In cross-section these valleys, when fully developed, are shallow, and dominated at a distance of one hundred to one hundred and fifty yards by the neighbouring heights. Apart from the occasional shrubs which align the watercourses this prairie resembles any beautiful Alpine meadow, for its high dense grass is in August strewn with countless flowers of various kinds, similar to those which one finds on lower levels in the mountains of Europe. This is what I shall call the upper world as opposed to the world of the large valleys. There are practically no permanent habitations on it. The herdsmen who live on it, wrongly described as nomads, practise a sort of seasonal migration, a change of abode, the extent of which is never more than a few score miles. They ascend to the higher pasture-lands in proportion as the temperature becomes more clement, and descend again with the temperature, remaining for a few weeks at each of their successive encampments. At the end of August the yak-herds[1] reach their highest grazing-lands at an altitude of 14,700 feet; then, towards the beginning of September, they descend towards the lower valleys where, at an altitude of 12,500 to 12,800 feet they spend the period of extreme cold, from November to April or May.

Well below the plateau is another world, completely different, the world of wide, deep valleys. It begins below an altitude of 11,500 feet and for that reason is limited to the narrow channels of the large rivers. This is the case in the valleys of Tao Fou and Luho, as well as in certain parts of the Ser Khio and Dohi valleys. The narrow fluvial terraces which border the rivers, near the cones of dejection of the torrents, are used as cultivated barley-fields and as soil for various types of vegetables, and these cultivated lands are sometimes artificially constructed terraces on the steeply inclined slopes of the young valleys, in the classic V-formation. So, whereas at a height of 14,700 feet, on the plateau, one would imagine oneself on some gently undulating plain, in the valleys, 3,300 feet lower down, one might be in the Alps. The illusion is completed by the patches of

[1] There are also herds of goats. Sheep are rare in the districts we visited. We were told that further to the north great numbers of them are to be found.

fir forest which form dark green stains against the lighter green of the slopes. There are also trees near the villages, at the entrances to the secondary valleys, trees of temperate regions, some of which reach an imposing size. It is at the bottom of these narrow valleys formed by the great rivers, in places where they are not constricted into impassable gorges, that all the sedentary life of the country is concentrated. The houses, grouped in hamlets or small villages, are of stone and square-shaped, of fortress-like dimensions, with low doors and occasional slit windows pierced only in the upper floors to guard against attacks which are inevitable in this country so often disturbed by tribal warfare, and are topped by immense lofts in which the harvest is dried. It is unusual to travel more than three or four miles without finding a monastery of several hundred lamas, erected usually on mountain spurs, in sites chosen both for their beauty and for their strategic position.

The world of herdsmen and the world of settlers, that of the plateaux and that of the wide valleys communicate by means of the secondary valleys of the tributary streams. This is a sort of "no man's land" composed of narrow, almost impenetrable gorges, often cluttered up with virgin forests of the Alpine type, in which the trunks of dead trees, lying on the ground, hinder the progress of the caravans. Here, the process of rejuvenation by headward erosion is apparent everywhere and explains how the plateau is attacked bit by bit, and how the morphology of the country as a whole is rejuvenated. Travellers are even more wary of these gorges, in which their caravans of pack-animals are exposed to attack, than of the high passes.

The valley world occupies a small proportion of the total surface and yet it is more thickly populated than the world of the plateaux. In spite of difficulties of contact the two worlds are in some degree connected. Each world is an economic supplement to the other. The herdsmen bring cattle produce to the people of the valleys (butter, meat, leather, yak-skins, wool, etc.) and receive in exchange tea, tsampa and sometimes industrial products. These negotiations are to a great extent controlled by the monasteries, which constitute the only recognized economic power; being large landowners, these religious institutions do not scorn to practice banking and to stock merchandise. The indisputable moral authority which the lamas exercise on the pious populations both of the plateaux and of the valleys, favours their speculations. In this interdependence of herdsmen and settlers the latter get the best of the bargain. If need be they could dispense with what the former bring them, whereas the herdsmen could not exist without tea and tsampa. It is the same all the world over; the settlers, more advanced and more farseeing, have acquired from the domiciliary nature of their life the taste for hoarding and, encouraged in this habit by the example of the religious communities who grant them their support, they benefit by their superiority.

The poor people of the plateaux have on their side only the fear which they inspire. As is the case among almost all pastoral peoples the men are more or less lazy. It is the women who do nearly all the work; they milk the animals, prepare the food, and weave the yak-skins. The children watch over the herds and the old men swing the leather bottles in which the butter is churned. The adult males are riders and, if need be, warriors. It devolves on them to protect the camps, but since this task only occupies a small part of their energies, they tend to unite and form raiding-parties, which their moral code allows. I do not believe, however, that there are many professional bandits. In certain circumstances anyone can become a bandit, which does not mean that he is necessarily a felon.[1] Sometimes these plundering expeditions are extortion reprisals against the people of the valleys, but they are mostly directed against the caravans of strangers. The salt caravans from the Kou-Kou-Nor are apparently spared.

The lamaic doctrine forbids murder, and therefore the influence of the monasteries tempers slightly the brutal habits of these people, who by nature tend to put a low value on human life. But they have managed to escape from the Chinese temporal suzerainty of the province of Si-kang and do not recognize any central authority.

The contrast between these two worlds, which are geographically so close and yet so different, will appear, in my opinion, still more marked as soon as it is possible to start deep ethnological studies. Apart from the obvious differences in their mode of life it is probable that anthropological observations will reveal important morphological differences between the inhabitants of the plateaux and of the valleys.

[1] On this point I do not share the opinion of G. de Roerich who denies the existence of the Ngolos as an ethnical group. "They are," he says, "a band of felons, outlaws, who have taken refuge in this district and who are apparently still skulking there to escape punishment." The fact that in north-eastern Tibet the word "Ngolo" has become synonymous with "bandit" does not prove anything.

HYDROLOGY

The Sources of the Tong

THE partial failure of our expedition prevented me from completely solving the problem of the sources of the Tong. Nevertheless, I located two main branches of this river, and can, without too much guesswork, deduce its upper ramifications. The theory adopted on various maps, according to which the drainage of this part of the Ngolo country is credited to the Yalong, must be abandoned. The upper reaches of the river which the Filchner expedition discovered to the north of the 33rd parallel clearly do not fit the Yalong but the Tong. The Ngolo-Seta country apparently lies almost entirely in the upper basin of the Tong. The district inhabited by these semi-nomad herdsmen coincides so well with this river's drainage basin that one must inevitably suppose a sort of territorial apportionment, the tribes who compose the Ngolo group apparently regarding as their demesne all the territories whose waters flow towards the Tong upstream from Damba.

Of the two large rivers which give birth to the Tong the one to the north, called the Dohi,[1] which I discovered after the disaster of the expedition is certainly the stronger. I was astonished to discover its importance several hundred miles upstream from Ouassekou where I last glimpsed the Tong. It is no mountain torrent but a real broad, unfordable river, with a rapid flow and swirling waters, its bed sometimes nearly a hundred yards wide, with no rocks visible, and whose fall, as far as I could ascertain in cross-section, shows no abrupt changes. Although I had no altimeter with me I guessed it to be 14,800 feet in altitude, and I think this estimate is pretty accurate. At the place where I left it it appeared to be flowing into deep gorges, and its fall was obviously tending to increase.

In my opinion, this river Dohi is the main arm of the Tong. Judging by the volume of its discharge, it is reasonable to suppose that its course, upstream from the point where I discovered it, extends roughly sixty miles. It is, therefore, very difficult to reconcile it with the headwaters discovered by Filchner, which are quite close, unless these are in reality situated farther to the west than Filchner indicated. But, allowing that the longitudes determined by the

[1] It is also known as the Ngolo-Ma-Tchou.

German expedition were wrong, it would be too much to suppose that the error would amount to between forty or fifty minutes. Yet, if allowance were made for this difference towards the west, it would reconcile Filchner's headwaters to our Dohi. It is more likely, however, that our river has its source a little farther westwards of the 100° meridian, towards the 33rd parallel, in that part of the Bayen-Khara mountains which we saw from the Liotard pass, at the foot of that imposing massif, crowned with perpetual snows and probably with glaciers which top them. It is probably due to this veritable reservoir that the Tong becomes an important river at the moment of its first appearance.

The Ser Khio, the other arm of the Tong which we discovered, is distinctly less important, and may be regarded as a tributary of the Dohi. This river, too, must have its source in the Bayen-Khara mountains, but one cannot fairly attribute more than 40 miles length to its course upstream from Chortaintong. It seems therefore impossible to make it conform with Filchner's sources.

The Dohi and the Ser unite probably towards the 101st meridian. Nothing is known about the tributaries which add their flow to the single river thus formed. If Filchner's sources are correctly placed, it is to their streams that one must attribute the drainage of the districts on the left bank of the upper basin of the Tong.

Fed by all these streams the Dohi is certainly an important river at its junction with the river Matang. Probably its discharge is even greater than the discharge of the Matang. Nevertheless it is the Matang which is usually shown on the maps as the main upper arm of the Tong. I think that I can explain this geographical error from the fact that, so far as I know, no explorer has seen the river junction. The few travellers who, coming from Sseu-tchouan, have reached the Tong by the valley of the river Matang, or by the valley of one of its tributaries, have all deviated from the junction and have crossed the mountains over the pass of Tan Chao Wou, at an altitude of 14,100 feet, whereas the bottom of the valley is probably 8,200 feet in altitude. This is what H. G. Thompson did in his 1925 journey from the Tong to the Houang-Ho. For several hours these travellers lost sight of the river which they had followed, and naturally supposed that the river they found farther on was the same, never suspecting the existence of a junction with another and more important water-course. No doubt the Tong at this point flows through a deeply entrenched gorge, difficult to penetrate, which forces the trail to deviate from the valley. Whatever the explanation, this fact has been the cause of a great many errors in the mapping of this district. Thompson, moreover, admits it implicitly when he writes: "At Damba I took a course towards the north-east, crossing the Tan Chao Wou pass at 14,130 feet, and reached the Taking (Tong) River, here called the Matang River, at Cherh Tsz." He had certainly reached the Matang, but he had left the Tong, of which the Matang and the Pou Pieng Ho are no doubt the largest left bank tributaries.

So the question of the Tong sources seems now to have been solved, although a great deal still remains to be discovered. The geographical importance of this strong river which drains mountain systems comprising some of the highest summits in the world (Mynia Gonka and Jara) is further increased by the fact that it takes its source somewhere in the Bayen-Kharas, massifs which reach 19,700 feet in altitude.

Delimitation of the upper basin of the Tong, in particular near the Yalong

By identifying two important upper arms of the Tong, our expedition has partly delimited the basin of that river in the country of the Ngolo-Setas. On the one hand our journey from Tatsienlou to Luho led us along the southern slopes of the high mountain chain which constitute for a certain distance the watershed of the rivers Tong and Yalong; it is therefore interesting to make an adjustment which enables one to delimit the basins of these rivers, from the massif of Mynia Gonka as far as the district of Kandze, that is to say about 150 miles.

In regard to the Mynia Gonka Mr. H. L. Richardson was kind enough to let me have a copy of the map which he drew up in accordance with his own observations and those of various travellers, notably Burdsall and Emmons, Heim and Edgar. It appears quite clearly on this map, when transferred to mine without change, that the whole massif of Mynia Gonka and all the peaks around Tatsienlou are drained by the Tong. Only a few slender streams reach the tributaries of the Yalong. The mastery of the Tong is evidently facilitated by the fact that it flows right along the foot of the enormous massif, whose melted snows on its eastern slope flow directly into the Tong's main valley. The distance as the crow flies between the summits of Mynia Gonka and the river is hardly twenty miles, whereas the difference in level is over 19,700 feet. One can imagine the impetus of the torrent which descends this slope. The Yalong, in the same latitude, is fed only from an altitude of 9,800 feet,[1] and therefore cannot rival its young and fiery competitor.

The drainage of the Mynia Gonka massif shows clearly that the line of the highest peaks does not always coincide with that of the watershed.

To the north of the 30th parallel things become more normal. The chain of peaks of over 16,700 feet altitude, which, starting from

[1] 9,188 feet at Hokeou.

the pass of Tcheto (or Sila), extend on both sides of the Jara nearly ninety-five miles to the north-west is indeed an ideal line which divides the two basins. The two slopes do not communicate except by passes of over 17,800 feet altitude, and therefore present no problem up to the northernmost summit of the Tao mountains; here an exact investigation would have to be made to decide whether a few unimportant little streams belong to one basin or the other.

To the north-west of the Tao mountains the general lowering of the relief makes the watershed much more confused. Nevertheless it preserves quite clearly the general north-west orientation of the chain of the Tao mountains. The Ser Khio, the tributary of the Tong, tries twice to cross this chain, but is thrown back each time to the north-west. At the place where we crossed this chain, from the basin of the Yalong to that of the Tong, we found, over a stretch of three or four miles, a gently undulating plateau, almost a peneplain, on which, through very open valleys separated by moderate hills, wind thin streams of water over a marshy soil. These channels appear at first doubtful which way to go; then, feeling their tremendous head, they plunge into V-shaped valleys in which they become wild torrents. Squeezed between the tributaries of the Tong and of the Yalong, the narrow remains of the plateau is everywhere attacked by the general rejuvenation of the relief. But here the Yalong wins, because, at the same longitude, its tributary the Seh has sawn its valley two hundred yards deeper than the upper arm of the Tong, the Ser Khio.

The district where the Seh, the Gni and the Ser Khio take their sources is still unknown.

GEOGRAPHICAL CO-ORDINATES OF A FEW LOCALITIES

	Louis Liotard's Observations	*Observations of Tan and Li—National Geographical Society of China*
Tatsienlou	30° 01′ 26″ N (south of the town) 101° 51′ 20″ E	30° 02′ 55″ N 101° 57′ 19″ E (of Greenwich)
Tao Fou	30° 58′ 25″ N (south of the town) 101° 09′ E	30° 58′ 52″ N 101° 07′ 41″ E (of Greenwich)
Luho	31° 23′ 43″ N No observations	31° 23′ 07″ N 100° 39′ 30″ E (of Greenwich)

ALTITUDES OF CERTAIN LOCALITIES

Tatsienlou . .	8,620 feet	(Mean of several observations with the boiling-point apparatus).
Tcheto (or Sila) Pass	14,106 „	Altimeter.
Taining (Gata) .	11,970 „	„
Solingkai Pass .	13,035 „	„
Tao . . .	10,890 „	(Mean of several observations with the boiling-point apparatus).
Peak of Tao . .	15,550 „	Altimeter
Luho . . .	11,950 „	(Observations with the boiling-point apparatus made near the monastery.)
Serba (Chinese post)	11,800 „	(Observation with the boiling-point apparatus.)

INDEX OF TIBETAN AND CHINESE WORDS

ÇAKTI	Source of feminine energy in the Lamaic mythology, which takes the form of a female demon.
ÇAKYA-MOUNI	Name most often given to the founder of Buddhism, Gautama, the Buddha.
ÇIVA	Third person of the Hindu Trinity, god of destruction, of healing and of fruitfulness.
MONT CHAKPORI	Situated in the south-western suburb of Lhassa. Here is founded the College of Medicine of Tibet.
CHORTAIN	Tibetan name for the Buddhist stupa, a kind of monument in the shape of a bulbed tower which originally served as a shrine and which is very common in Tibet.
CHAMBHALA	Legendary country which lies somewhere in the north, closed and inaccessible.
DALAÏ-LAMA	Chief reincarnation of Tibet, who at the same time exerts temporal power.
DZO	Product of the cross-breeding of the domestic cow and the yak.
DORDJE	Tibetan name for the vajra of the Hindu mythology: fulminant stone, thunderbolt; the object which represents it is the symbol of power, and plays an important part in the Lama ritual.
DRIL-BU	Little bronze bell topped by a half-dordje, which also plays a great part in the ritual.
GAOU	Metal or leather talisman which the Tibetans carry about with them.
GELUKPA	Name of a Lama sect which has at its head the Dalaï-Lama.
KANAN DRO	Where are you going?
KALE, KALE	Gently, Gently. Equivalent of "Goodbye".
KESAR	Epic hero of Central Asia.
KOMBA	House.

Index of Tibetan and Chinese Words

KHATA (or scarf of happiness)	White muslin scarf used as an offering and as an accompaniment for an exchange of civilities; equivalent of a visiting-card.
MANI	Religious monument made by heaping up stones with religious inscriptions engraved on them.
MAÏTREYA	Future Buddha of the Lama mythology.
OM MANI PADME HOUM	"O jewel in the lotus!" Sacred formula of Lamaism, used constantly by the lamas and laymen, and inscribed everywhere on the monuments, the prayer-wheels, the manis, etc.
PILING	Stranger.
POUKHAÏ	Chinese blanket or covering into which one wraps oneself for sleeping.
PADMA SAMBHAVA	The most famous magician of the Lama mythology.
SATIA	Lama sect of an earlier date than the Gelukpa sect.
SWASTIKA	Religious symbol.
TANKHA	Religious painting on canvas which hangs from the ceilings or walls of Lama buildings.
TCHAPA	Brigand.
TSAMPA	Barley meal lightly roasted, which is the Tibetan's basic food.
TCHOUBA	Big Tibetan cloak.
TSONG KAPA	Reformer of Lamaism in the fourteenth century and founder of the Gelukpa sect.
YIN and YANG	Personification of the male and female in the mythologies of the Far East.
YA PO RE	It's good; it's well.
YAMEN	Official residence of a Chinese civil servant.

INDEX

Index

Index

Index

Some other Oxford Paperbacks for readers interested in Central Asia, China and South-east Asia, past and present

CAMBODIA

GEORGE COEDÈS
Angkor

MALCOLM MacDONALD
Ankor and the Khmers*

CENTRAL ASIA

PETER FLEMING
Bayonets to Lhasa

ANDRÉ GUIBAUT
Tibetan Venture

LADY MACARTNEY
An English Lady in Chinese Turkestan

DIANA SHIPTON
The Antique Land

C. P. SKRINE AND PAMELA NIGHTINGALE
Macartney at Kashgar*

ALBERT VON LE COQ
Buried Treasures of Chinese Turkestan

AITCHEN K. WU
Turkistan Tumult

CHINA

All About Shanghai: A Standard Guide

HAROLD ACTON
Peonies and Ponies

VICKI BAUM
Shanghai '37

ERNEST BRAMAH
Kai Lung's Golden Hours*

ERNEST BRAMAH
The Wallet of Kai Lung*

ANN BRIDGE
The Ginger Griffin

CHANG HSIN-HAI
The Fabulous Concubine*

CARL CROW
Handbook for China

PETER FLEMING
The Siege at Peking

MARY HOOKER
Behind the Scenes in Peking

CORRINNE LAMB
The Chinese Festive Board

W. SOMERSET MAUGHAM
On a Chinese Screen*

G. E. MORRISON
An Australian in China

DESMOND NEILL
Elegant Flower

PETER QUENNELL
A Superficial Journey through Tokyo and Peking

OSBERT SITWELL
Escape with Me! An Oriental Sketch-book

J. A. TURNER
Kwang Tung or Five Years in South China

HONG KONG AND MACAU

AUSTIN COATES
City of Broken Promises

AUSTIN COATES
A Macao Narrative

AUSTIN COATES
Myself a Mandarin

AUSTIN COATES
The Road

The Hong Kong Guide 1893

INDONESIA

S. TAKDIR ALISJAHBANA
Indonesia: Social and Cultural Revolution

DAVID ATTENBOROUGH
Zoo Quest for a Dragon*

VICKI BAUM
A Tale from Bali*

'BENGAL CIVILIAN'
Rambles in Java and the Straits in 1852

MIGUEL COVARRUBIAS
Island of Bali*

BERYL DE ZOETE AND WALTER SPIES
Dance and Drama in Bali

AUGUSTA DE WIT
Java: Facts and Fancies

JACQUES DUMARÇAY
Borobudur

JACQUES DUMARÇAY
The Temples of Java

ANNA FORBES
Unbeaten Tracks in Islands of the Far East

GEOFFREY GORER
Bali and Angkor

JENNIFER LINDSAY
Javanese Gamelan

EDWIN M. LOEB
Sumatra: Its History and People

MOCHTAR LUBIS
The Outlaw and Other Stories

MOCHTAR LUBIS
Twilight in Djakarta

MADELON H. LULOFS
Coolie*

MADELON H. LULOFS
Rubber

COLIN McPHEE
A House in Bali*

ERIC MJOBERG
Forest Life and Adventures in the Malay Archipelago

HICKMAN POWELL
The Last Paradise

E. R. SCIDMORE
Java, Garden of the East

MICHAEL SMITHIES
Yogyakarta: Cultural Heart of Indonesia

LADISLAO SZÉKELY
Tropic Fever: The Adventures of a Planter in Sumatra

EDWARD C. VAN NESS AND SHITA PRAWIROHARDJO
Javanese Wayang Kulit

MALAYSIA

ISABELLA L. BIRD
The Golden Chersonese: Travels in Malaya in 1879

MARGARET BROOKE THE RANEE OF SARAWAK
My Life in Sarawak

HENRI FAUCONNIER
The Soul of Malaya

W. R. GEDDES
Nine Dayak Nights

A. G. GLENISTER
The Birds of the Malay

Peninsula, Singapore and Penang

C. W. HARRISON
Illustrated Guide to the Federated Malay States (1923)

BARBARA HARRISSON
Orang-Utan

TOM HARRISSON
World Within: A Borneo Story

CHARLES HOSE
The Field-Book of a Jungle-Wallah

EMILY INNES
The Chersonese with the Gilding Off

W. SOMERSET MAUGHAM
Ah King and Other Stories*

W. SOMERSET MAUGHAM
The Casuarina Tree*

MARY McMINNIES
The Flying Fox*

ROBERT PAYNE
The White Rajahs of Sarawak

OWEN RUTTER
The Pirate Wind

ROBERT W. SHELFORD
A Naturalist in Borneo

CARVETH WELLS
Six Years in the Malay Jungle

SINGAPORE

RUSSELL GRENFELL
Main Fleet to Singapore

R. W. E. HARPER AND HARRY MILLER
Singapore Mutiny

JANET LIM
Sold for Silver

G. M. REITH
Handbook to Singapore (1907)

C. E. WURTZBURG
Raffles of the Eastern Isles

THAILAND

CARL BOCK
Temples and Elephants

REGINALD CAMPBELL
Teak-Wallah

MALCOLM SMITH
A Physician at the Court of Siam

ERNEST YOUNG
The Kingdom of the Yellow Robe

Titles marked with an asterisk have restricted rights.